State Estimation and Control for Low-cost Unmanned Aerial Vehicles

Chingiz Hajiyev • Halil Ersin Soken
Sıtkı Yenal Vural

State Estimation and Control for Low-cost Unmanned Aerial Vehicles

Springer

Chingiz Hajiyev
Faculty of Aeronautics and Astronautics
Istanbul Technical University
Istanbul, Turkey

Sıtkı Yenal Vural
Faculty of Aeronautics and Astronautics
Istanbul Technical University
Istanbul, Turkey

Halil Ersin Soken
Institute of Space and Astronautical Science (ISAS)
Japan Aerospace Exploration Agency (JAXA)
Sagamihara, Kanagawa, Japan

ISBN 978-3-319-16416-8 ISBN 978-3-319-16417-5 (eBook)
DOI 10.1007/978-3-319-16417-5

Library of Congress Control Number: 2015941902

Springer Cham Heidelberg New York Dordrecht London

Printed on acid-free paper

Springer International Publishing AG Switzerland is part of Springer Science+Business Media (www.springer.com)

Preface

Over time, technological developments in aeronautics have a bigger impact in all aspects of human life and, surely, unmanned aerial vehicles (UAVs) play an important role in this process.

The main aim of this book is to make a contribution to progress in UAV technology, especially in the sense of autonomous control, and that is achieved in three stages: fault-free estimation of UAV parameters by robust adaptive Kalman filters, using fault detection and isolation for the integration of UAV flight data, and controlling the UAV by means of classical or fuzzy controllers.

The state estimation, control, and fault detection and isolation methods for low-cost UAVs are illustrated by examples and computer simulations performed via the software package MATLAB.

Most of the book is original and contributed by the authors.

This book is a research monograph in which the theory and application of the proposed methods have been discussed together.

This monograph is composed of 11 chapters, as follows:

Chapter 1 is the introduction to the book. An introduction to UAVs along with a brief history is given. It is discussed why UAVs are important and what are their advantages over conventional manned planes

In **Chap. 2**, equations of motion for UAVs are given. The coordinate systems, derivation of rigid body equations of motion, and linearized equations of motion for UAVs are presented. The state-space representation of the linearized system is also included in this chapter.

In **Chap. 3**, navigation systems for UAVs are given. In this context, inertial navigation, air data and satellite radio navigation systems, Doppler and altimeter radars, and magnetic sensors are described. Moreover, measurement faults for these systems, when they are used in UAVs, are investigated and details for the fault modeling are presented.

In **Chap. 4**, the optimal linear Kalman filter (OKF) algorithm is presented. The OKF for UAV state estimation is shown. The stability of the OKF and necessity of Kalman filter adaptation are discussed.

In **Chap. 5**, a robust Kalman filter (RKF) with filter gain correction for the case of measurement malfunctions is introduced. By the use of a defined variable named the measurement noise scale factor (MNSF), the faulty measurements are taken into the consideration with a small weight and the estimations are corrected without affecting the characteristics of the accurate measurements. The RKF algorithms with single and multiple MNSFs (*R*-adaptation procedure) are proposed and applied for the state estimation process of the UAV platform in the presence of measurement faults. The results of these algorithms are compared for different types of sensor faults and recommendations about their utilization are given. Stability analysis for proposed RKFs is also included.

In **Chap. 6**, the process noise covariance adaptation procedure (*Q*-adaptation) for UAV state estimation is introduced and the robust adaptive Kalman filter (RAKF) algorithm with the *R*- and *Q*-adaptations against sensor/actuator failures is proposed. Thus, the filter stands robust against the faults and, even in case of sensor/actuator failure, continues providing accurate estimation results. The performance of the proposed RAKF is investigated by simulations for the state estimation procedure of UAVs. The presented RAKF ensures that the parameter estimation system of the UAV is not influenced by sensor/actuator failures and, therefore, autonomous missions can be performed successfully.

In **Chap. 7**, Global Positioning System (GPS)/Inertial Navigation System (INS) measurements, which have high accuracy, and air data system (ADS) measurements, which have low accuracy but high frequency, are integrated using the Kalman filtering technique in order to obtain high-accuracy measurement data at high frequency. It is shown that the designed system, which is based on the indirect Kalman filtering technique, is successful in finding the wind speed data using the known error dynamics of the system and the determined statistical values. The fault detection and isolation (FDI) algorithms are developed by using the proposed system and diagnostic tests are performed in order to evaluate the performance of the system under different sensor measurement fault scenarios. For fusing the data coming from different measurement sets, a federated Kalman filter is used. The best data are produced by using the FDI algorithms and leaving the faulty data out before fusing with the federated filter.

The stability of UAVs is investigated in **Chap. 8** for the purpose of controller design. The transfer functions for the longitudinal and lateral stability analysis of UAVs are introduced and discussed in this chapter.

The design procedure for the flight control system of UAVs based on the classical controllers is presented in **Chap. 9**. Classical control methods, such as the proportional-integral-derivative (PID) controllers, are investigated. It is observed that the PID controller responses are good at controlling the longitudinal and lateral flight dynamics of small UAVs.

In **Chap. 10**, an optimal controller is designed using the linear quadratic regulator (LQR) method. In order to better evaluate the effect of disturbances on the obtained measurements, a Kalman filter is used. Initially, the controller is tested for a case where disturbances are absent. Then, a Kalman filter is designed and the system under disturbances is tested with the designed controller and filter. The results reveal the effectiveness of the Kalman filter and LQR controller.

In **Chap. 11**, a fuzzy controller for UAV flight dynamics is investigated. Fuzzy logic-based longitudinal and lateral controllers are designed. The fuzzy logic controller works well, even though no optimization technique is used to optimize the system, and it was designed without dynamic model knowledge. In fuzzy logic controllers, a careful arrangement of membership functions may lead to better results. The stability of the fuzzy controllers is investigated. The effectiveness of different control methods for UAV flight control is compared in this chapter.

This monograph is useful for both researchers in academia and professional engineers in the aerospace industry. It is also an important reference book for graduate research students. Moreover, Chaps. 2, 3, 4, 8, 9, and 10 can also be used by undergraduate students studying aerospace and control and systems engineering.

We would like to thank Mr. Volkan Kargin for his contribution to the simulations given in Chap. 7 and M.Sc. student Demet Cilden who helped us with the corrections and formatting of the book.

We hope that this book will be helpful to researchers, students, our colleagues, and all the readers.

Istanbul, Turkey
January, 2015

Chingiz Hajiyev
Halil Ersin Soken
Sıtkı Yenal Vural

Contents

Chapter 1
Introduction to Unmanned Aerial Vehicles

1.1 Introduction

Everything started with a passion and a desire for flying like birds. Was looking to the world from above only a dream for humanity? Maybe that was the only thought of Hezarfen Ahmet Celebi when he jumped down from Galata Tower four centuries ago, for his first flight over Istanbul. Ahmet Celebi is a legendary character in Turkish history and we will probably never be sure of his flight's reality, but there is one thing that we are certain of; much has changed since that date. Flight technology has come to a point that he could not imagine, if he ever existed. Today, aircraft are ideal vehicles of transportation, perfect weapons of warfare, and helpers of human beings, which makes life easier in many areas. Now, think that they are not very expensive or hard to produce and do not even need a pilot. UAVs, namely, unmanned aerial vehicles, were developed for making this purpose real. Although they do not have a genius director, a human aboard, for controlling extraordinary situations without risking human life, UAVs are preferred to manned aircraft for their properties of being more economical, smaller, and lighter.

From the beginning, as can be understood from its name, a UAV is a vehicle without a pilot or crew aboard, and when a battle operation, risky mission, or a special case like redirecting a guided missile is discussed, that specialty makes them greatly superior. For years, wars have been a potential death machine for humanity and, in that situation, having an assistant which can reduce your army's loss of life becomes very important. As Holder states, the availability of using them during dangerous missions, when rapid information is needed, manned aircraft cannot be operated, and in other situations that they are required, gives UAVs an important role in warfare [1]. Supporting that, possibly the best example of the usage of UAVs in battle is the Iraq war. The eyes and assistants of the ground-based coalition forces in civil war areas of Iraq were UAVs [2]. If the Fallujah operation, a hard, key operation of the war, is examined, that truth can be seen.West explains the

C. Hajiyev et al., *State Estimation and Control for Low-cost Unmanned Aerial Vehicles*, DOI 10.1007/978-3-319-16417-5_1

position before the operation; since the consultation between the USA and Iraqi governments finished, they agreed on an operation to Fallujah for taking control of that part of Baghdad again. The American and Iraqi military forces, comprising 12,000 persons, must fight with the 4,000 rebels without a serious loss of life, which threatened both them and about 200,000 civilians in the city. At that point, finding the location of the rebels became important and the only unit of the army which could do that perfectly was Marine Unmanned Aerial Squadron (VMU-1), nicknamed "Watchdogs", and their Pioneer UAVs [2]. At the end of the operation, after 11 days, the city was cleared and the result of the mission was nearly 30 times lower life loss for multinational forces. This precisely demonstrates how useful UAVs can be in a war scenario.

Another situation that suits a UAV is risky operations. Perhaps the attacking capability of a UAV is limited, but their excellence in discovery missions, which are described as the most risky operations, cannot be questioned [1]. For example, Newcome declares that a life-threatening type of job for pilots is carrying out discovery flights in nuclear test areas, which occurred with high possibility during the Cold War, the result of which could bring about a nuclear war immediately. For that reason, these missions were described as being "dirty", so giving the job to UAVs was the best approach [3]. They really are excellent in risky missions and, from this point of view, Holder aids Newcome as he describes another advantage of a UAV; there is no need for a risky rescue mission with the potential for more loss of life when a UAV is lost [1]. In 2000, after the volcanic activity of Mount Usu in Japan, instead of manned aircraft, UAVs flew over the area by the decision of the Ministry of Construction. The reason for using UAVs for the discovery of dangerous situations was the active volcano, deemed too risky for human pilots [3]. Also, there have been some strange suggestions made as a result of the suitability of UAVs in "dirty" duties. For instance, according to Holder, making UAVs a clear target for guided missiles and not causing the loss of a pilot's life is one of the other ideas which is examined for the usage of UAVs [1]. When the most valuable thing of a pilot, his or her life, is in danger, UAVs are their lifebuoys.

In today's world, it is a truth that making an artificial bird fly is very expensive and, because of that, more economical UAVs may be attractive. Firstly, the education process of a pilot is very long and, related to that, not cheap, so operating a UAV for the same jobs can be better for economy, simply because it does not need a pilot. A combat pilot candidate can learn everything about fighting and flying in 6 weeks, a period nearly 17 times longer than the duration of the building process of a new Predator UAV [3]. There is no need to waste time and money. Also, they have the possibility of being used as part of a pilots' fighting skill development process. That secures both a better as well as a cheaper education. Being a moveable target for fighter aircraft weapons is a popular use for drone aircrafts. Holder defines as an example and, he continues, for educating their fighter pilots and improving their combat talent, the Air Force and Navy of the US Army have used various kinds of drone aircrafts for years.

Another direct economical advantage of UAV usage is the price of a UAV. When the verification between manned and unmanned aircraft is done in a security area, this fact can be seen more clearly. In 2005, Farivar noted that aircraft used by police-like organizations for surveillance and crime-fighting missions demanded a lot of money to buy, replace, and repair [4]. Hence, more economical UAVs that can replace manned planes became more interesting. Again, according to Farivar, the person responsible for UAV technology development for Los Angeles County Sheriff's Department, Cmdr. Sid Heal, explains "the plane's low cost makes it highly desirable, particularly when compared with the current alternative of a helicopter" and, he continues, "that helicopter costs $450 an hour—for every 10 h of use I've bought a new one of these that I own." Also, he claims "they shoot one of these down and I'll have funds for 100 of them the next day" [4]. There is one thing that is certain; people want aircraft which are cheap on account to both their construction and flight costs, and that is valid even for armies with large budgets.

Thirdly, being unmanned reduces the size of an aircraft and, when compared with large planes, a small plane has many more advantages. Being unmanned does not only mean not having a pilot; it also means not carrying many more devices necessary for the crew. For example, life support equipment is an essential part of manned aircraft. Holder indicates that, when manned aerial craft and UAVs are considered, not requiring life support equipment for a pilot puts UAVs a step ahead [1]. That equipment is big and heavy, so if the same conditions are under discussion, using a small UAV with less chance of being spotted is more reasonable; nobody wants his or her spy plane to be seen. Some other advantages that small size brings to a plane are long flight times and ease of transportation and taking off. As a result, UAVs are considered a part of many missions, such as fire fighting. DeGaspari tells his readers that, unlike the limited vision of ground-based fire fighting personnel, aerial vehicles have the chance to explore wide fire areas. For meeting that requirement, the laborers of the Idaho National Engineering and Environmental Laboratory (INEEL) are working on a fire fighter UAV [5]. Scott Bauer, program manager for the lab's UAV program at the INEEL, says that carrying a small UAV with a trailer, sending it airborne with the help of a catapult, and controlling it with a suitcase-sized remote control is plausible, where he describes a small UAV as a vehicle with 3–3.5-m wingspan, 12-kg payload capacity, and 10 h of flight time [5]. At the same time, there are other examples of UAVs which are easier to transport because of their portability and, surprisingly, they can be ready to fly in a few minutes. Security UAVs, which are used by law-keeping organizations like the police, are one of these kinds of vehicles. According to Farivar regarding the size of a security UAV, Dr. Chang explains that the plane is portable enough to carry in a bag and, when it is required for an operation, mounting it and making it fly in a short time is possible [4].

Making a heavy object fly is always harder than flying a lighter one and, in that case, the property of an UAV being lighter can be examined. First, for purposes like security keeping, traffic controlling, and having vision for the press, operating a UAV over cities is safer. Farivar explains the situation by using a quotation from Prof. Kroo, the security UAVs director of the Aircraft Aerodynamics and Design Group at Stanford University. Prof. Ilan Kroo states that, "the miniaturization of these aircrafts is also driven by a concern for public safety" and he adds, "if something catastrophic happens with the airplane and it crashes in a populated region, you want this thing to weigh a few ounces, not a few thousand pounds" [4]. At the same time, the advantages of lower mass go hand in hand with being small. For example, a small and light UAV can be hand-launched. A javelin commercial UAV with the ability of hand launching, a property related to its 6.8-kg weight, has wide usage potential, such as live vision from the air for the press and law enforcement organizations [1]. Another thing, lightness affects their invisibility. Without the need for crew, the weight of the aircraft decreases nearly to its lower limit, and that is something greatly desired by aerial early warning (AEW) system developers because a light vehicle has greater opportunity to fly higher and for long enough without being seen [6]. Lighter is always better.

In conclusion, being unmanned, more economical, smaller, and lighter makes UAVs superior to manned craft; at that point, not having a genius director no longer becomes important. The thing which is fascinating is that the visibility of UAVs' supremacy is increasing day by day. Already, in many areas, UAVs have taken the place of manned aircraft. Who knows, maybe one day, even passenger planes will be UAVs and there will no longer be a need for pilots.

1.2 UAV Types and Applications

UAVs can be defined as aircraft without a human pilot aboard. According to the International Civil Aviation Organization (ICAO), unmanned aircraft is classified into two groups [7, 8]:

(A) *Autonomous aircraft*

Autonomous aircraft are considered not to be suitable for regulation due to legal and liability issues

(B) *Remotely piloted aircraft*

Remotely piloted aircraft are subject to civil regulation under the ICAO and under the relevant national aviation authority

There are also different names given to UAVs. They are called unmanned air systems, unpiloted air systems, drones, unpiloted aerial vehicles, etc. The flight of UAVs can be controlled using on-board computers autonomously (following the online calculations or precoded control algorithms) or by remote control by a pilot on the ground. UAV systems are complex systems that comprise:

- Control station used by system operators
- The UAV itself
- Communications system

UAVs can be used for many different military and civilian applications. These include:

Civilian uses:

- Aerial photography
- Agricultural uses
- Meteorological purposes
- Traffic monitoring
- Logistics: UAVs for cargo operations
- Police authority
- Fire services
- Information services
- Research and development: as an experiment for further studies

Military uses:

- Reconnaissance
- Surveillance
- Target and decoy: functioning as an enemy missile or aircraft for aerial or ground gunnery
- Combat: as fighter or bomber aircraft (in general) on high-risk missions
- Electronic intelligence

UAVs can be classified into groups depending on their size and capability for specific missions. Endurance, range, weight, and maximum altitude characteristics determine the capability of UAVs, and they are used to perform the classification. However, these definitions are constantly changing as the technology advances and new types of UAVs emerge. Different types of UAVs in use today are [7, 8]:

- Orbital UAVs – UAVs that may operate up to low Earth orbit. They can reach speeds of Mach 25+ and a primary mission example is launching spacecraft into orbit.
- Hypersonic UAVs – UAVs that may reach supersonic (Mach 1–5) or hypersonic (Mach 5+) speeds. They are suitable for flying at suborbital altitude.
- HALE (high altitude long endurance) – They operate over 15,000 m altitude and for a period of 24 h or more. They are usually operated by the military.
- MALE (medium altitude long endurance) – They are operated at altitudes between 5,000 and 15,000 m. They are generally used by the military.
- Tactical UAVs – Used by the military at smaller ranges. They are simpler systems compared to MALE and HALE.
- Close-range UAVs – Used by mobile army groups. Their ranges are up to 100 km. They are mostly used for surveillance, traffic monitoring, crop spraying, and reconnaissance missions. Handheld UAVs may be regarded as a subcategory.
- Mini-UAVs – UAVs of up to 20 kg approximately. They may be hand-launched. They are usually used by mobile army groups.
- Micro-UAVs – UAVs having a wingspan no greater than 150 mm. They are required for operation in urban environments.
- Nano air vehicles – These are proposed to be on the same size scale as insects. They can be used for ultrashort-range surveillance.

The listed types of aircraft can be either rotary wing aircraft (which are able to land and take off vertically) or fixed-wing aircraft.

UAVs can also be classified according to their wing loading, engine type, and power loading.

For calculating the wing loading, the total weight of the UAV is divided by the wing area. Three classes have been created to classify UAVs according their wing loading values. UAVs that have a wing loading value greater than 100 kg/m^2 are classified as having high loading. Those that have a wing loading value less than 100 kg/m^2 but more than 50 kg/m^2 are classified as having medium wing loading. The other UAVs with wing loading values less than 50 kg/m^2 are classified as having low wing loading.

To complete different types of missions, different engines are used in UAVs. These include turbofans, two-stroke engines, piston, rotary, turboprop, electric, and propeller. Some examples of UAVs that use different engines are given below:

- Global Hawk – military UAV –turbofan
- Predator – military UAV – piston
- Dragon Eye – military UAV – electric

1.3 A Brief History of UAVs

"Unmanned aerial vehicle" is a term which came into general use in the early 1990s and replaced the previously preferred expression of "remotely piloted vehicle" (RPV) [3]. According to Newcome [3], a UAV is:

> A powered, aerial vehicle that does not carry a human operator, uses aerodynamic forces to provide vehicle lift, can fly autonomously or be piloted remotely, can be expendable or recoverable, and can carry a lethal or nonlethal payload. Ballistic or semiballistic vehicles, cruise missiles, and artillery projectiles are not considered unmanned aerial vehicles.

Although UAVs is a recent concept which has become popular following uses especially in NATO operations during the Kosovo War and American operations like Operation Enduring Freedom, it has a history as old as manned planes from World War I. In fact, misunderstanding their role in military missions and not seeing their accomplishments clearly are key reasons why military planners did not think of them as a part of operations earnestly until the mid-1980s [9], which brought about a reduced public interest in those vehicles up to that date. However, nowadays, UAVs are used in many public applications, such as traffic monitoring, fire fighting, etc., as well as their crucial function in military missions, like reconnaissance, attack, etc.

Historically, the "Kettering Bug" built by Charles Kettering for the USA Navy in 1918 may be accepted as the first UAV. This prototype was successful enough to make the army order additional ones, but it has never flown operationally because of the termination of World War I. On the other hand, in the light of common opinion, it is possible to show the "Lightning Bug", which was substantially used by the USA during the Vietnam War between 1964 and 1972, as the ancestor of modern UAVs.

During that period, more than 3,400 combat UAVs were flown over North Vietnam, Laos, and China by Strategic Air Command 100th Strategic Reconnaissance Wing of the USAF [9].

In the near future, it is expected that UAV technology will make progress dramatically. If this development phase is examined from the point of view of this study, then certainly, it is required to be the focus of researches about the autonomy level of UAVs. The performance of autonomous missions, where planes have the ability to make decisions without human intervention, is directly affected by wrong decisions and not so precise positioning, attitude determination, etc. In this sense, first, an accurate method for the estimation of UAV states and, then, an efficient algorithm for controlling these states are what we need for increasing the performance of autonomous UAV missions. Several algorithms presented in this book aim to solve these problems.

1.4 Conclusion

As stated, the properties of being unmanned, more economical, smaller, and lighter make UAVs superior to manned planes and, at that point, not having a genius director no longer becomes important. The thing which is fascinating is that the visibility of UAVs' supremacy is increasing day by day. Already, in many areas, UAVs have taken the place of manned aircraft. Maybe one day, even passenger planes will be UAVs and there will no longer be a need for pilots. However, a crucial challenge awaiting us before we can use UAVs in all areas of human life is to increase the autonomy level of these vehicles, leading to a minimum of human interaction. Several algorithms introduced in this book aim to solve this issue.

References

1. Holder WG, Holder B (2001) Unmanned air vehicles: an illustrated study of UAVs. Schiffer Publishing, Atglen
2. West B (2005) Nowhere to hide. Pop Mech 182(2):54–59
3. Newcome LR (2004) Unmanned aviation: a brief history of unmanned aerial vehicles. AIAA, Reston
4. Farivar C (2005) A flying crime fighter (some assembly required). New York Times (Article dated: January, 13, 2005), p. G7
5. DeGaspari J (2005) Unmanned fire spies [Electronic version]. Mech Eng 127(2):20–22
6. Long MW (ed) (1992) Airborne early warning system concepts. Artech House Publishing, Norwood
7. Austin R (2010) Unmanned aircraft systems: UAVs design, development and deployment. Wiley, Chichester
8. Fahlstrom PG, Gleason TJ (2012) Introduction to UAV systems, 4th edn. Wiley, Hoboken
9. Cook KLB (2007) The silent force multiplier: the history and role of UAVs in warfare. IEEE aerospace conference, Big Sky, MT, USA, March 2007

Chapter 2
Equations of Motion for an Unmanned Aerial Vehicle

2.1 Rigid Body Equations of Motion

Before going into deeper analyses of the equations of motion, it will be useful to review the axis systems that will be dealt with in further steps.

2.1.1 Coordinate Systems

2.1.1.1 Aircraft Body Axis System

The origin of the frame is located at the center of gravity of the aircraft. The x-axis is directed toward the nose of the plane, while the y-axis is defined as coming out of its right wing. The z-axis completes the system by the right-hand rule as it points down through the bottom of the aircraft.

2.1.1.2 The Earth Axis System

The z-axis of the system shows the center of the Earth. The system lies on the local horizontal plane of the Earth, as the orthogonal x- and y-axes are defined in the directions of north and east, respectively. As a general manner, the Earth axis system is assumed to be the inertial axis for aircraft-related studies. Since aircraft rotation rates are relatively high in comparison to the rotation rate of the Earth, this assumption gives acceptably low error for aircraft problems and can be utilized [1, 2] (Fig. 2.1).

C. Hajiyev et al., *State Estimation and Control for Low-cost Unmanned Aerial Vehicles*, DOI 10.1007/978-3-319-16417-5_2

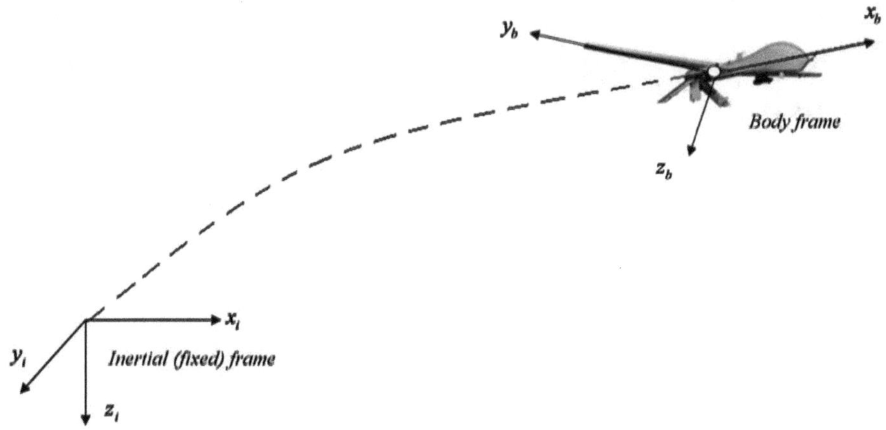

Fig. 2.1 Body and inertial axis system

2.1.2 *Derivation of Rigid Body Equations of Motion*

Under the assumptions that the aircraft is a rigid body and its mass remains constant for a relatively short duration of time, Newton's second law can be written as:

$$\vec{F} = m\left(\frac{d\vec{V}_c}{dt}\right)_i \tag{2.1}$$

where m is the mass of the aircraft, V_c is the velocity of the center of mass of the airplane, $\vec{F}$ is the force acting on the plane, and subscript i indicates that the derivation is taken with respect to the inertial frame.

If the equation is rewritten in order to fix the axis system to the aircraft body frame, rotation of the plane with respect to the inertial frame must be taken into consideration. The derivative of $\vec{V}_c$, referred to a rotating body frame that has an angular velocity of $\vec{\omega}_b$, can be shown by:

$$\left(\frac{d\vec{V}_c}{dt}\right)_i = \left(\frac{d\vec{V}_c}{dt}\right)_b + \vec{\omega}_b \, x \left(\vec{V}_c\right)_b \tag{2.2}$$

Hence, $\vec{F}$, the force that acts on the aircraft, becomes:

$$\vec{F} = m\left[\left(\frac{d\vec{V}_c}{dt}\right)_b + \vec{\omega}_b \, x \left(\vec{V}_c\right)_b\right] \tag{2.3}$$

where $\left(\vec{V_c}\right)_b$, the velocity of the aircraft in the body frame, is formed by u, v, and w, velocities in the x, y, and z directions, respectively, as:

$$\left(\vec{V_c}\right)_b = u\vec{i} + v\vec{j} + w\vec{k} \tag{2.4}$$

and $\vec{\omega}_b$, the angular velocity vector in the body frame, consists of p, r, and q, angular rates about the x, y, and z axes (or roll, pitch, and yaw rates), respectively, as:

$$\vec{\omega}_b = p\vec{i} + r\vec{j} + q\vec{k} \tag{2.5}$$

As a result,the three force equations for the aircraft in the body axis may be given as:

$$\begin{bmatrix} F_x \\ F_y \\ F_z \end{bmatrix} \equiv m \begin{bmatrix} \dot{u} \\ \dot{v} \\ \dot{w} \end{bmatrix} + m \begin{vmatrix} \vec{i} & \vec{j} & \vec{k} \\ p & q & r \\ u & v & w \end{vmatrix} = m \begin{bmatrix} \dot{u} + qw - rv \\ \dot{v} + ru - pw \\ \dot{w} + pv - qu \end{bmatrix} \tag{2.6a}$$

$$\begin{aligned} F_x &= m\,(\dot{u} + qw - rv) \\ F_y &= m\,(\dot{v} + ru - pw) \\ F_z &= m\,(\dot{w} + pv - qu) \end{aligned} \tag{2.6b}$$

On the other hand, in order to derive the three moment equations of the aircraft, a more complicated methodology may be followed based on examining an arbitrary small elemental mass, δm, which is at a distance from the center of mass of the aircraft and rotates about it [2] (Fig. 2.2).

Fig. 2.2 Arbitrary small differential mass in the body frame

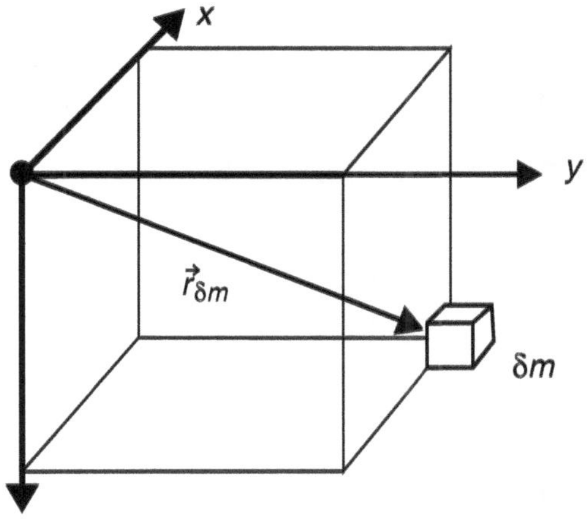

First, the distance between the differential mass and the center of mass of the aircraft is given by the following vector:

$$\vec{r}_{\delta m} = x\vec{i} + y\vec{j} + z\vec{k} \tag{2.7}$$

Here, x, y, and z are the distances in the x, y, and z axes. Now, the velocity of that differential mass with respect to the center of mass should be written as:

$$\vec{V}_{\delta m} = \left(\frac{\mathrm{d}\vec{r}_{\delta m}}{\mathrm{d}t}\right)_b + \vec{\omega}_b\, x\, \vec{r}_{\delta m} \tag{2.8}$$

Since the aircraft is assumed to be rigid, $\vec{r}_{\delta m}$ is constant in time and, so:

$$\left(\frac{\mathrm{d}\vec{r}_{\delta m}}{\mathrm{d}t}\right)_b = 0 \tag{2.9}$$

Therefore:

$$\vec{V}_{\delta m} = \vec{\omega}_b\, x\, \vec{r}_{\delta m} = \begin{vmatrix} \vec{i} & \vec{j} & \vec{k} \\ p & q & r \\ x & y & z \end{vmatrix} \tag{2.10a}$$

$$\Rightarrow \vec{V}_{\delta m} = (qz - ry)\,\vec{i} + (rx - pz)\,\vec{j} + (py - qx)\,\vec{k} \tag{2.10b}$$

Furthermore, an expression for the angular momentum of differential mass, $\delta\vec{H}_{\delta m}$, can be derived as:

$$\delta\vec{H}_{\delta m} = \vec{r}_{\delta m}\, x \left(\delta m \vec{V}_{\delta m}\right) \tag{2.11}$$

Hence, the three terms of the angular momentum can be obtained as:

$$\delta\vec{H}_{\delta m} = \begin{bmatrix} \delta H_x \\ \delta H_y \\ \delta H_z \end{bmatrix} = \begin{vmatrix} \vec{i} & \vec{j} & \vec{k} \\ x & y & z \\ \delta m\,(qz - ry) & \delta m\,(rx - pz) & \delta m\,(py - qx) \end{vmatrix} \tag{2.12a}$$

$$\begin{aligned} & \delta H_x = p\left(y^2 + z^2\right)\delta m - qxy\,\delta m - rxz\,\delta m \\ \Rightarrow\ & \delta H_y = q\left(x^2 + z^2\right)\delta m - ryz\,\delta m - pxy\,\delta m \\ & \delta H_z = r\left(x^2 + y^2\right)\delta m - pxz\,\delta m - qyz\,\delta m \end{aligned} \tag{2.12b}$$

After that, the expressions should be integrated over the whole aircraft to determine the angular momentum equations of plane itself. As long as p, r, and q are not dependent on mass, then:

$$
\begin{aligned}
H_x &= \iiint \delta H_x = p\iiint (y^2+z^2)\,\delta m - q\iiint xy\,\delta m - r\iiint xz\,\delta m \\
H_y &= \iiint \delta H_y = q\iiint (x^2+z^2)\,\delta m - r\iiint yz\,\delta m - p\iiint xy\,\delta m \\
H_z &= \iiint \delta H_z = r\iiint (x^2+y^2)\,\delta m - p\iiint xz\,\delta m - q\iiint yz\,\delta m
\end{aligned}
\tag{2.13}
$$

It is known that moments of inertia are defined as:

$$I_{xx} = \iiint (y^2+z^2)\,\delta m \tag{2.14a}$$

$$I_{yy} = \iiint (x^2+z^2)\,\delta m \tag{2.14b}$$

$$I_{zz} = \iiint (x^2+y^2)\,\delta m \tag{2.14c}$$

and the products of inertia are:

$$I_{xy} = \iiint xy\delta m \tag{2.15a}$$

$$I_{yz} = \iiint yz\delta m \tag{2.15b}$$

$$I_{xz} = \iiint xz\delta m \tag{2.15c}$$

Consequently, if these moments of inertia and products of inertia terms are substituted into Eq. (2.13), then:

$$
\begin{aligned}
H_x &= pI_{xx} - qI_{xy} - rI_{xz} \\
H_y &= qI_{yy} - rI_{yz} - pI_{xy} \\
H_z &= rI_{zz} - pI_{xz} - qI_{yz}
\end{aligned}
\tag{2.16a}
$$

or in matrix notation:

$$
\begin{bmatrix} H_x \\ H_y \\ H_z \end{bmatrix} = \begin{bmatrix} I_{xx} & -I_{xy} & -I_{xz} \\ -I_{xy} & I_{yy} & -I_{yz} \\ -I_{xz} & -I_{yz} & I_{zz} \end{bmatrix} \begin{bmatrix} p \\ q \\ r \end{bmatrix}
\tag{2.16b}
$$

In general, the x–z plane of the aircraft is assumed to be a plane of symmetry and, so:

$$I_{xy} = I_{yz} = 0 \tag{2.17}$$

This simplifies the two parts of Eq. (2.16) and:

$$\begin{aligned} H_x &= pI_{xx} - rI_{xz} \\ H_y &= qI_{yy} \\ H_z &= rI_{zz} - pI_{xz} \end{aligned} \tag{2.18}$$

As the final stage, the time derivative of the angular momentum vector should be determined. In a similar way to Eq. (2.2), the time derivative of $\vec{H}$, referred to a rotating body frame that has an angular velocity of $\vec{\omega}_b$, can be expressed as:

$$\left(\frac{d\vec{H}}{dt}\right)_i = \left(\frac{d\vec{H}}{dt}\right)_b + \vec{\omega}_b \, x \, \vec{H}_b \tag{2.19}$$

Since it is assumed that the mass and the mass distribution of the plane are constant, the time moments of inertias and the products of inertias do not vary with time:

$$\dot{I}_{xx} = \dot{I}_{yy} = \dot{I}_{zz} = \dot{I}_{xz} = 0 \tag{2.20}$$

Apart from that point:

$$\left(\frac{d\vec{H}}{dt}\right)_b = \begin{bmatrix} \dot{p}I_{xx} - \dot{r}I_{xz} \\ \dot{q}I_{yy} \\ \dot{r}I_{zz} - \dot{p}I_{xz} \end{bmatrix} \tag{2.21}$$

and it is obvious that:

$$\vec{\omega}_b \, x \, \vec{H}_b = \begin{vmatrix} \vec{i} & \vec{j} & \vec{k} \\ p & q & r \\ pI_{xx} - rI_{xz} & qI_{yy} & rI_{zz} - pI_{xz} \end{vmatrix} = \begin{bmatrix} q\left(rI_{zz} - pI_{xz}\right) - rqI_{yy} \\ r\left(pI_{xx} - rI_{xz}\right) - p\left(rI_{zz} - pI_{xz}\right) \\ pqI_{yy} - q\left(pI_{xx} - rI_{xz}\right) \end{bmatrix} \tag{2.22}$$

These yields:

$$\left(\frac{d\vec{H}}{dt}\right)_i = \left(\frac{d\vec{H}}{dt}\right)_b + \vec{\omega}_b \, x \, \vec{H}_b = \begin{bmatrix} \dot{p}I_{xx} + qr\left(I_{zz} - I_{yy}\right) - \left(\dot{r} + pq\right)I_{xz} \\ \dot{q}I_{yy} - pr\left(I_{zz} - I_{xx}\right) + \left(p^2 - r^2\right)I_{xz} \\ \dot{r}I_{zz} + pq\left(I_{yy} - I_{xx}\right) + \left(qr + \dot{p}\right)I_{xz} \end{bmatrix} \tag{2.23}$$

As a result, the three moment equations of the motion of the aircraft in the body axis system can be written as:

$$\begin{aligned} L &= \dot{p}I_{xx} + qr\left(I_{zz} - I_{yy}\right) - (\dot{r} + pq)\, I_{xz} \\ M &= \dot{q}I_{yy} - pr\,(I_{zz} - I_{xx}) + \left(p^2 - r^2\right) I_{xz} \\ N &= \dot{r}I_{zz} + pq\left(I_{yy} - I_{xx}\right) + (qr + \dot{p})\, I_{xz} \end{aligned} \tag{2.24}$$

Here, L is the moment around the x-axis, M is the moment around the y–axis, and N is the moment around the z-axis. Also, it is good to emphasize that the first terms of all three equation, $\dot{p}I_{xx}, \dot{q}I_{yy}, \dot{r}I_{zz}$, are related to the angular acceleration, while the second terms, $qr\left(I_{zz} - I_{yy}\right), pr\left(I_{zz} - I_{xx}\right), pq\left(I_{yy} - I_{xx}\right)$, denote gyro precision, and the third expressions, $\left(\dot{r} + pq\right) I_{xz}, \left(p^2 - r^2\right) I_{xz}, \left(qr + \dot{p}\right) I_{xz}$, are coupling terms.

Therefore, in summary, the three force and three moment equations for an aircraft are:

$$\text{Force Eq.} \quad \Rightarrow \quad \begin{aligned} F_x &= m\,(\dot{u} + qw - rv) \\ F_y &= m\,(\dot{v} + ru - pw) \\ F_z &= m\,(\dot{w} + pv - qu) \end{aligned}$$

$$\text{Moment Eq.} \quad \Rightarrow \quad \begin{aligned} L &= \dot{p}I_{xx} + qr\left(I_{zz} - I_{yy}\right) - (\dot{r} + pq)\, I_{xz} \\ M &= \dot{q}I_{yy} - pr\,(I_{zz} - I_{xx}) + \left(p^2 - r^2\right) I_{xz} \\ N &= \dot{r}I_{zz} + pq\left(I_{yy} - I_{xx}\right) + (qr + \dot{p})\, I_{xz} \end{aligned}$$

2.2 Orientation and Position of an Aircraft

The equations of motion of an aircraft examined thus far are derived for the body frame of the plane. However, the orientation and position of the aircraft cannot be defined in a relatively moving frame and, instead, a fixed frame such as the Earth axis system must be used.

Hence, it is necessary to introduce some methodology to relate these two axis systems: the fixed inertial frame and the aircraft body frame. To achieve this, the orientation of the aircraft can be described by three consecutive rotations, and these angular rotations are called Euler angles [1], a concept developed by Leonhard Euler to define the orientation of a rigid body in 3D Euclidean space.

For such a problem, the order of rotations is important and the 3-2-1 Euler angle representation is used. That means:

- First, rotate about the z-axis through the yaw angle ψ
- Then, as the second step, rotate about the y-axis through the pitch angle θ
- And, finally, rotate about the z-axis through the roll angle of ϕ

As a consequence, it is possible to associate flight velocity components relative to the fixed reference frame and the velocity components in the body frame:

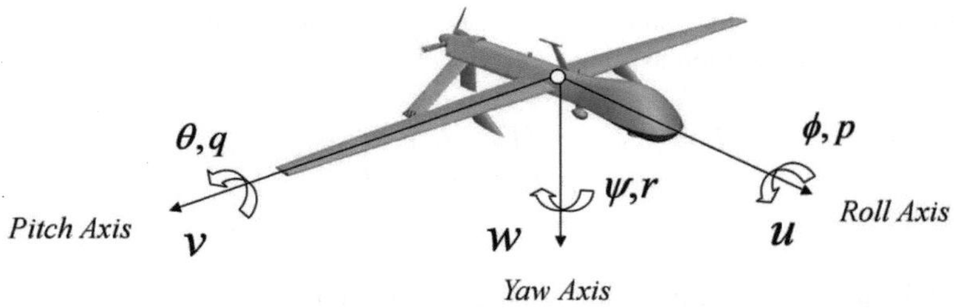

Fig. 2.3 Euler angles, velocity components, and angular velocities for an unmanned aerial vehicle (UAV)

$$\begin{bmatrix} \frac{dx}{dt} \\ \frac{dy}{dt} \\ \frac{dz}{dt} \end{bmatrix} = \begin{bmatrix} c\theta c\psi & s\phi s\theta c\psi - c\phi s\psi & c\phi s\theta c\psi - s\phi s\psi \\ c\theta s\psi & s\phi s\theta s\psi + c\phi c\psi & c\phi s\theta s\psi - s\phi c\psi \\ -s\theta & s\theta c\theta & c\phi c\theta \end{bmatrix} \begin{bmatrix} u \\ v \\ w \end{bmatrix} \tag{2.25}$$

Here, u, v, and w are velocity components in the body frame, $\frac{dx}{dt}$, $\frac{dy}{dt}$, and $\frac{dz}{dt}$ are velocity components in the fixed inertial frame, and θ, ψ, and ϕ are Euler angles. $c(\cdot)$ and $s(\cdot)$ are cosine and sine functions, respectively.

Per contra, by the use of Euler angles, angular rates (or angular velocities) of p, q, and r may be related with the Euler rates of $\dot{\theta}$, $\dot{\psi}$, and $\dot{\phi}$ (Fig. 2.3):

$$\begin{bmatrix} p \\ q \\ r \end{bmatrix} = \begin{bmatrix} 1 & 0 & -s\theta \\ 0 & c\phi & c\theta s\phi \\ 0 & -s\phi & c\theta c\phi \end{bmatrix} \begin{bmatrix} \dot{\phi} \\ \dot{\theta} \\ \dot{\psi} \end{bmatrix} \tag{2.26a}$$

or, inversely:

$$\begin{bmatrix} \dot{\phi} \\ \dot{\theta} \\ \dot{\psi} \end{bmatrix} = \begin{bmatrix} 1 & s\phi \tan\theta & c\phi \tan\theta \\ 0 & c\phi & -s\phi \\ 0 & s\phi \sec\theta & c\phi \sec\theta \end{bmatrix} \begin{bmatrix} p \\ q \\ r \end{bmatrix} \tag{2.26b}$$

2.3 Small Perturbation Theory

Small perturbation theory is a technique for linearizing nonlinear equations. By performing this mathematical method, it is possible to find approximate solutions to problems that cannot be solved exactly.

As long as the six equations of motion of the plane, as derived in the previous section, are nonlinear and must be linearized for solving behavior problems of aircraft, small perturbation theory must be used to progress. First of all, in order to apply the method, it is assumed that the motion of the aircraft consists of small perturbations about the steady flight condition [1]. Then, as a base for the theory, small perturbation terms are added to the steady states:

$$\begin{aligned} u &= u_0 + \Delta u \quad p = p_0 + \Delta p \quad \theta = \theta_0 + \Delta\theta \\ v &= v_0 + \Delta v \quad q = q_0 + \Delta \mathrm{q} \quad \phi = \phi_0 + \Delta\phi \\ w &= w_0 + \Delta w \quad r = r_0 + \Delta r \quad \psi = \psi_0 + \Delta\psi \end{aligned} \tag{2.27}$$

If the reference flight condition is assumed to be symmetric, the propulsive forces are assumed to remain constant and the x-axis is considered, as it coincides with the direction of the aircraft's velocity vector, then the perturbed states become:

$$\begin{aligned} u &= u_0 + \Delta u \quad p = \Delta p \; \theta = \theta_0 + \Delta\theta \\ v &= \Delta v \quad q = \Delta \mathrm{q} \quad \phi = \Delta\phi \\ w &= \Delta w \quad r = \Delta r \quad \psi = \Delta\psi \end{aligned} \tag{2.28}$$

Here, u, v, and w are velocity components in the x, y, and z directions, p, q, and r are angular velocities about the x, y, and z axes, and θ, ϕ, and ψ are the pitch, roll, and yaw angles. The subscript 0 indicates the steady states, where Δ is used for perturbations.

Now, let's handle the force equation in the x-direction as an example:

$$F_x = X = m\left(\dot{u} + qw - rv\right) \tag{2.29}$$

Replace all of the states with perturbed ones:

$$X_0 + \Delta X = m\left(\Delta\dot{u} + \Delta q \Delta w - \Delta r \Delta v\right) \tag{2.30}$$

At level flight conditions, there is no acceleration and, so, $X_0 = 0$. On the other hand, if order of $(\Delta q), (\Delta w), (\Delta r), (\Delta v) = \varepsilon$, neglect terms with the order of ε^2. Hence:

$$\Delta X = m\left(\Delta\dot{u}\right) \tag{2.31}$$

If the same method is followed for all six equations of motion, linearized equations may be obtained as follows:

$$\frac{1}{m}\begin{bmatrix} \Delta X \\ \Delta Y \\ \Delta Z \end{bmatrix} = \begin{bmatrix} \Delta\dot{u} \\ \Delta\dot{v} - \Delta r u_0 \\ \Delta\dot{w} - \Delta q u_0 \end{bmatrix} \tag{2.32}$$

$$\begin{bmatrix} \Delta L \\ \Delta M \\ \Delta N \end{bmatrix} = \begin{bmatrix} I_{xx}\dot{p} + I_{zz}\dot{r} \\ I_{yy}\dot{q} \\ I_{zz}\dot{r} + I_{xz}\dot{p} \end{bmatrix} \tag{2.33}$$

Besides, ΔX can be expressed by means of a Taylor series in terms of the perturbation variables as:

$$\Delta X = \frac{\partial X}{\partial u}\Delta u + \frac{\partial X}{\partial w}\Delta w + \frac{\partial X}{\partial \delta_e}\Delta \delta_e + \frac{\partial X}{\partial \delta_T}\Delta \delta_T \tag{2.34}$$

where $\Delta\delta_e$, $\Delta\delta_T$ are control-related terms and stand for the elevator deflection and change in thrust, respectively. Then, these two equations can be related as follows:

$$m\left(\Delta\dot{u}\right) = \frac{\partial X}{\partial u}\Delta u + \frac{\partial X}{\partial w}\Delta w + \frac{\partial X}{\partial \delta_e}\Delta \delta_e + \frac{\partial X}{\partial \delta_T}\Delta \delta_T \tag{2.35}$$

If the same kind of association is built for all six linearized equations and, after that, necessary simplifications are realized under the light of assumptions, the result will be the longitudinal and lateral equations of motions for the aircraft, which is also a consequence of small perturbation theory. To learn more about the technique, [1] and [2] are good references.

2.4 Linearized Equations of Motion

In this final part of the chapter, first, linearized equations of motion of an airplane are given in state space form and, then, by taking the specifications of the Zagi unmanned aerial vehicle (UAV) [3] as a guideline, the equations are rewritten. Zagi is a flying wing type radio-controlled aircraft that is popular amongst hobbyists.

At this point, it might be useful to indicate that we mainly take references [3–5] as an essence to obtain scalar quantities of stability derivatives and, therefore, stability coefficients. Hence, the values are accepted as true and used as they are in general. Some small modifications are applied, like changing the initial conditions, if necessary.

2.4.1 Equations in General

Linearized longitudinal equations of motion of the plane can be written in state space form as:

$$\begin{bmatrix} \Delta\dot{u} \\ \Delta\dot{w} \\ \Delta\dot{q} \\ \Delta\dot{\theta} \end{bmatrix} = \begin{bmatrix} X_u & X_w & 0 & -g \\ Z_u & Z_w & u_0 & 0 \\ M_u + M_{\dot{w}}Z_u & M_w + M_{\dot{w}}Z_w & M_q + M_w u_0 & 0 \\ 0 & 0 & 1 & 0 \end{bmatrix} \begin{bmatrix} \Delta u \\ \Delta w \\ \Delta q \\ \Delta\theta \end{bmatrix} + \begin{bmatrix} X_{\delta e} & X_{\delta T} \\ Z_{\delta e} & Z_{\delta T} \\ M_{\delta e} + M_{\dot{w}}Z_{\delta e} & M_{\delta T} + M_{\dot{w}}Z_{\delta T} \\ 0 & 0 \end{bmatrix} \begin{bmatrix} \Delta\delta_e \\ \Delta\delta_T \end{bmatrix} \tag{2.36a}$$

Following this, lateral equations of motion can be given as:

$$\begin{bmatrix} \Delta\dot{v} \\ \Delta\dot{p} \\ \Delta\dot{r} \\ \Delta\dot{\phi} \end{bmatrix} = \begin{bmatrix} Y_v & Y_p & -(u_0 - Y_r) & g\cos(\theta_0) \\ L_v & L_p & L_r & 0 \\ N_v & N_p & N_r & 0 \\ 0 & 1 & 0 & 0 \end{bmatrix} \begin{bmatrix} \Delta v \\ \Delta p \\ \Delta r \\ \Delta\phi \end{bmatrix} + \begin{bmatrix} 0 & Y_{\delta r} \\ L_{\delta a} & L_{\delta r} \\ N_{\delta a} & N_{\delta r} \\ 0 & 0 \end{bmatrix} \begin{bmatrix} \Delta\delta_a \\ \Delta\delta_r \end{bmatrix} \tag{2.36b}$$

Here, $\Delta\delta_e$, $\Delta\delta_a$, and $\Delta\delta_r$ are elevator, aileron, and rudder deflections, respectively, $\Delta\delta_T$ is the change in thrust, and $X_u, Z_u \ldots Z_{\delta e} \ldots Y_v, L_v \ldots L_{\delta a} \ldots$ are stability derivatives.

2.4.2 *Characteristics of the Chosen Zagi UAV*

As an experimental platform, the Zagi UAV is chosen and, in the remainder of the book, the proposed methods are demonstrated using the dynamics and characteristics of this UAV (Fig. 2.4).

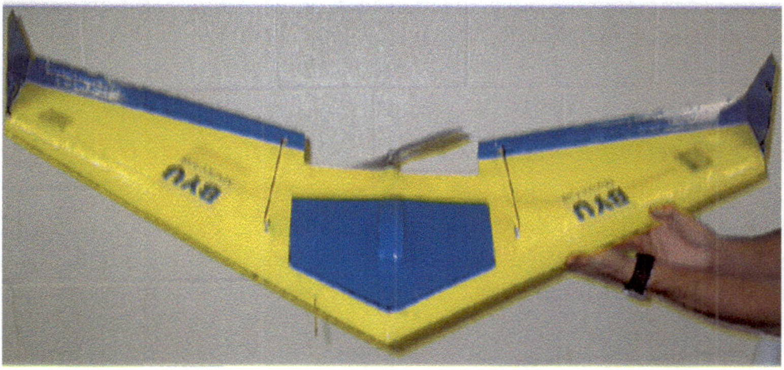

Fig. 2.4 The Zagi UAV used by the students of Brigham Young University [4]

Table 2.1 Specifications of the Zagi unmanned aerial vehicle (UAV) [3]

m	1.56 kg
S	0.2589 m^2
b	1.4224 m
$\bar{c}$	0.3302 m
$J =$	$\begin{pmatrix} 0.1147 & 0 & -0.0015 \\ 0 & 0.0576 & 0 \\ -0.0015 & 0 & 0.1712 \end{pmatrix}$ kg.m^2

Table 2.2 Stability coefficients of the Zagi UAV [3]

Longitudinal coefficients	Lateral coefficients
$C_{L_0} = 0.28$	$C_{Y_{70}} = 0$
$C_{D_0} = 0.03$	$C_{l_0} = 0$
$C_{m_0} = 0$	$C_{n_0} = 0$
$C_{L_\alpha} = 3.45$	$C_{Y_\beta} = -0.98$
$C_{D_\alpha} = 0.30$	$C_{l_\beta} = -0.12$
$C_{m_\alpha} = -0.38$	$C_{n_\beta} = 0.25$
$C_{L_q} = 0$	$C_{Y_p} = 0$
$C_{D_q} = 0$	$C_{l_p} = -0.26$
$C_{m_q} = -3.6$	$C_{n_p} = 0.022$
$C_{L_{\delta_e}} = -0.36$	$C_{Y_r} = 0$
$C_{D_{\delta_e}} = 0$	$C_{l_r} = 0.14$
$C_{m_{\delta_e}} = 0.5$	$C_{n_r} = -0.35$
	$C_{Y_{\delta_a}} = 0$
	$C_{l_{\delta_a}} = 0.08$
	$C_{n_{\delta_a}} = 0.06$
	$C_{Y_{\delta_r}} = -0.17$
	$C_{l_{\delta_r}} = 0.105$
	$C_{n_{\delta_r}} = -0.032$

In Table 2.1, m is the mass, S is the wing area, b is the span length, $\bar{c}$ is the mean chord length, and J is the inertia matrix of the plane. Notwithstanding, the stability coefficients of the Zagi UAV, which are used in order to obtain stability derivatives, may be tabulated as in Table 2.1.

2.4.3 *Linearized Equations of Motion for the Zagi UAV*

In this last step, stability derivatives calculated for the Zagi UAV, in the light of Tables 2.1 and 2.2 and related equations from [5], are substituted in the linearized lateral and longitudinal equations of motion. However, before that process, the equations are modified in such a way that observing the height of the UAV is also possible and the sideslip angle is used instead of the starboard velocity v:

$$\begin{bmatrix} \Delta\dot{u} \\ \Delta w \\ \Delta\dot{q} \\ \Delta\dot{\theta} \\ \Delta\dot{h} \end{bmatrix} = \begin{bmatrix} X_u & X_w & 0 & -g & 0 \\ Z_u & Z_w & u_0 & 0 & 0 \\ M_u + M_{\dot{w}}Z_u & M_w + M_{\dot{w}}Z_w & M_q + M_w u_0 & 0 & 0 \\ 0 & 0 & 1 & 0 & 0 \\ 0 & -1 & 0 & u_0 & 0 \end{bmatrix} \begin{bmatrix} \Delta u \\ \Delta w \\ \Delta q \\ \Delta\theta \\ \Delta h \end{bmatrix}$$

$$+ \begin{bmatrix} X_{\delta e} & X_{\delta T} \\ Z_{\delta e} & Z_{\delta T} \\ M_{\delta e} + M_{\dot{w}}Z_{\delta e} & M_{\delta T} + M_{\dot{w}}Z_{\delta T} \\ 0 & 0 \\ 0 & 0 \end{bmatrix} \begin{bmatrix} \Delta\delta_e \\ \Delta\delta_T \end{bmatrix} \tag{2.37a}$$

$$\begin{bmatrix} \Delta\dot{\beta} \\ \Delta\dot{p} \\ \Delta\dot{r} \\ \Delta\dot{\phi} \end{bmatrix} = \begin{bmatrix} \frac{Y_\beta}{u_0} & \frac{Y_p}{u_0} & -\frac{u_0 - Y_r}{u_0} & \frac{g\cos(\theta_0)}{u_0} \\ L_\beta & L_p & L_r & 0 \\ N_\beta & N_p & N_r & 0 \\ 0 & 1 & 0 & 0 \end{bmatrix} \begin{bmatrix} \Delta\beta \\ \Delta w \\ \Delta q \\ \Delta\theta \end{bmatrix} + \begin{bmatrix} 0 & \frac{Y_{\delta r}}{u_0} \\ L_{\delta a} & L_{\delta r} \\ N_{\delta a} & N_{\delta r} \\ 0 & 0 \end{bmatrix} \begin{bmatrix} \Delta\delta_a \\ \Delta\delta_r \end{bmatrix} \tag{2.37b}$$

Here, h is the height of the plane and β is the sideslip angle.

Therefore, as the calculated quantities are substituted into those equations, specialized equations for the Zagi UAV will become:

$$\begin{bmatrix} \Delta\dot{u} \\ \Delta\dot{w} \\ \Delta\dot{q} \\ \Delta\dot{\theta} \\ \Delta\dot{h} \end{bmatrix} = \begin{bmatrix} -0.3356 & 1.3181 & 0 & -9.80665 & 0 \\ -1.7916 & -3.9003 & 9.8215 & 0 & 0 \\ 0.702 & -3.5375 & -11.392 & 0 & 0 \\ 0 & 0 & 1 & 0 & 0 \\ 0 & -1 & 0 & 9.8215 & 0 \end{bmatrix} \begin{bmatrix} \Delta u \\ \Delta w \\ \Delta q \\ \Delta\theta \\ \Delta h \end{bmatrix}$$

$$+ \begin{bmatrix} -0.7436 & 6.8728 \\ 3.7855 & 0 \\ 47.917 & 0 \\ 0 & 0 \\ 0 & 0 \end{bmatrix} \begin{bmatrix} \Delta\delta_e \\ \Delta\delta_T \end{bmatrix} \tag{2.38a}$$

$$\begin{bmatrix} \Delta\dot{\beta} \\ \Delta\dot{p} \\ \Delta\dot{r} \\ \Delta\dot{\phi} \end{bmatrix} = \begin{bmatrix} -0.1069 & 0.1962 & -1 & 0.9984 \\ -1.2213 & -1.9155 & 1.0096 & 0 \\ 1.7255 & 0.0919 & -1.7198 & 0 \\ 0 & 1 & 0 & 0 \end{bmatrix} \begin{bmatrix} \Delta\beta \\ \Delta w \\ \Delta q \\ \Delta\theta \end{bmatrix} + \begin{bmatrix} 0 & -0.1855 \\ 8.348 & 0 \\ 4.24 & -2.1272 \\ 0 & 0 \end{bmatrix} \begin{bmatrix} \Delta\delta_a \\ \Delta\delta_r \end{bmatrix} \tag{2.38b}$$

Note that it is also possible to show the equations in a joined way, as follows:

$$\begin{bmatrix} \Delta\dot{u} \\ \Delta\dot{w} \\ \Delta\dot{q} \\ \Delta\dot{\theta} \\ \Delta\dot{h} \\ \Delta\dot{\beta} \\ \Delta\dot{p} \\ \Delta\dot{r} \\ \Delta\dot{\phi} \end{bmatrix} = \begin{bmatrix} -0.3356 & 1.3181 & 0 & -9.80665 & 0 & 0 & 0 & 0 & 0 \\ -1.7916 & -3.9003 & 9.8215 & 0 & 0 & 0 & 0 & 0 & 0 \\ 0.702 & -3.5375 & -11.392 & 0 & 0 & 0 & 0 & 0 & 0 \\ 0 & 0 & 1 & 0 & 0 & 0 & 0 & 0 & 0 \\ 0 & -1 & 0 & 9.8215 & 0 & 0 & 0 & 0 & 0 \\ 0 & 0 & 0 & 0 & 0 & -0.1069 & 0.1962 & -1 & 0.9984 \\ 0 & 0 & 0 & 0 & 0 & -1.2213 & -1.9155 & 1.0096 & 0 \\ 0 & 0 & 0 & 0 & 0 & 1.7255 & 0.0919 & -1.7198 & 0 \\ 0 & 0 & 0 & 0 & 0 & 0 & 1 & 0 & 0 \end{bmatrix} + \begin{bmatrix} -0.7436 & 6.8728 & 0 & 0 \\ 3.7855 & 0 & 0 & 0 \\ 47.917 & 0 & 0 & 0 \\ 0 & 0 & 0 & 0 \\ 0 & 0 & 0 & 0 \\ 0 & 0 & 0 & -0.1855 \\ 0 & 0 & 8.348 & 0 \\ 0 & 0 & 4.24 & -2.1272 \\ 0 & 0 & 0 & 0 \end{bmatrix} \begin{bmatrix} \Delta\delta_e \\ \Delta\delta_T \\ \Delta\delta_a \\ \Delta\delta_r \end{bmatrix} \tag{2.39}$$

By regarding the general representation of the state space form as:

$$\dot{X} = AX + BU, \tag{2.40}$$

where A is the system matrix, B is the control distribution matrix, U is the control vector, and X is the state space vector, one can write the following:

$$A = \begin{bmatrix} -0.3356 & 1.3181 & 0 & -9.80665 & 0 & 0 & 0 & 0 & 0 \\ -1.7916 & -3.9003 & 9.8215 & 0 & 0 & 0 & 0 & 0 & 0 \\ 0.702 & -3.5375 & -11.392 & 0 & 0 & 0 & 0 & 0 & 0 \\ 0 & 0 & 1 & 0 & 0 & 0 & 0 & 0 & 0 \\ 0 & -1 & 0 & 9.8215 & 0 & 0 & 0 & 0 & 0 \\ 0 & 0 & 0 & 0 & 0 & -0.1069 & 0.1962 & -1 & 0.9984 \\ 0 & 0 & 0 & 0 & 0 & -1.2213 & -1.9155 & 1.0096 & 0 \\ 0 & 0 & 0 & 0 & 0 & 1.7255 & 0.0919 & -1.7198 & 0 \\ 0 & 0 & 0 & 0 & 0 & 0 & 1 & 0 & 0 \end{bmatrix} \tag{2.41a}$$

$$B = \begin{bmatrix} -0.7436 & 6.8728 & 0 & 0 \\ 3.7855 & 0 & 0 & 0 \\ 47.917 & 0 & 0 & 0 \\ 0 & 0 & 0 & 0 \\ 0 & 0 & 0 & 0 \\ 0 & 0 & 0 & -0.1855 \\ 0 & 0 & 8.348 & 0 \\ 0 & 0 & 4.24 & -2.1272 \\ 0 & 0 & 0 & 0 \end{bmatrix} \tag{2.41b}$$

$$U = \left[\Delta\delta_e \; \Delta\delta_T \; \Delta\delta_a \; \Delta\delta_r \right]^{\mathrm{T}} \tag{2.41c}$$

$$X = \left[\Delta u \; \Delta w \; \Delta q \; \Delta\theta \; \Delta h \; \Delta\beta \; \Delta p \; \Delta r \; \Delta\phi \right]^{\mathrm{T}} \tag{2.41d}$$

References

1. Nelson RC (1998) Flight stability and automatic control, 2nd edn. WCB/McGraw-Hill, Boston
2. Yechout TR (2003) Introduction to aircraft flight mechanics. AIAA Education Series, Reston
3. Matthews JS (2006) Adaptive control of micro air vehicles. M.Sc. thesis, Department of Electrical and Computer Engineering, Brigham Young University, Provo, UT, USA
4. Christiansen RS (2004) Design of an autopilot for small unmanned aerial vehicles. M.Sc. thesis, Brigham Young University, Provo, UT, USA
5. Vural Y (2007) Autopilot system design for a small unmanned aerial vehicle (in Turkish). M.Sc. thesis, Department of Aeronautical Engineering, Istanbul Technical University, Istanbul, Turkey

Chapter 3
Navigation Systems for Unmanned Aerial Vehicles

3.1 Two Main Categories in Navigation

It is possible to categorize navigation systems into two groups as follows [1]:

- Dead reckoning (DR) systems
- Positioning systems

The most developed version of navigation is the integration of two methods optimally via a Kalman filter.

DR navigation systems determine the position of a vehicle by estimating the range from a location using the velocity and direction data of the vehicle. They are self-sufficient and independent from external systems. Some DR systems used for aircraft are presented below:

- INS – inertial navigation system
- Doppler/attitude reference system
- Air data/attitude reference system

Positioning error increases with time for all DR navigation systems. Therefore, it is required to correct the DR positioning error and update the system with accurate positions from available positioning systems.

Positioning systems are navigation systems which generally locate on spacecraft or the Earth itself and have transmitters. An aircraft's receiver determines the position by using transmitted signals via an auxiliary computer. Nowadays, the Global Positioning System (GPS) is the most accurate and developed positioning system. Other positioning systems are GLONASS, VOR/DME, TACAN, etc.

C. Hajiyev et al., *State Estimation and Control for Low-cost Unmanned Aerial Vehicles*, DOI 10.1007/978-3-319-16417-5_3

3.2 Inertial Navigation

An INS determines the movement and position of a vehicle according to a reference inertial frame using the vehicle's inertial reactions.

An INS has trios of accelerometers and gyroscopes which are perpendicular to each other. Inertial navigation is done by measuring the acceleration of the vehicle and integrating it twice to find the velocity and position. Therefore, position can be found with respect to a specific center freely without any external signal and data. Accelerometers are used to find the magnitude of acceleration according to a specific reference coordinate system. Acceleration is a vector, so it has both magnitude and direction. Gyroscopes are used to determine the direction of acceleration. Gyroscopes are used to keep accelerometers on a specific coordinate system, called the inertial space. The ring angles, which are the angles between the gyroscope's rings, determine the Euler angles, which show the angular orientation of the vehicle. Thus, an INS provides both angular orientation and forward movement.

An INS has three main functions [2]:

1. Sensing
2. Computation
3. Output

Accelerometers and gyroscopes provide the sensing function. Acceleration and angular velocity data from them are sent to the computer, which computes the velocity, position, orientation or attitude, altitude, and range to the destination by using these data.

If the real air velocity is supplied by the air data system, the INS can calculate the velocity and direction of the wind. An output function sends the data to the flight control system, engine control system, target sensors, and control display unit with respect to features of the mission.

There are two types of INS:

1. Gimballed
2. Analytical (strapdown)

In gimballed systems (classical), accelerometers and gyroscopes are isolated against rotation maneuvers, i.e., angular movements of the aircraft. Therefore, INS devices stay constantly in a specific orientation according to the Earth or inertial space. This case simplifies velocity and position calculations, and reduces the requirement for gyroscopes, because, otherwise, gyroscopes have to measure great rotation ratios.

In strapdown, acceleration and Euler angles are measured with respect to the aircraft coordinate system, multiplied with a transfer matrix, and then transformed to the main coordinate system. This procedure requires very fast computer processing, so this system only started to come into use after the 1980s. These processes—angle measurement and coordinate system transformation—must be done quickly by the computer. The strapdown system does not require rings, engines, etc.

Typically, an INS is a self-contained system with high short-term stability and is not influenced by interference. An INS is a standalone navigational system using motion sensors to continuously keep track of the position, orientation, and velocity of a vehicle [3].

The main advantages and disadvantages of an INS are as follows [3]:

Advantages of an INS:

An INS has many advantages in comparison with other navigation systems:

1. Position and velocity data are sudden and permanent. High data ratio and bandwidth are obtained easily.
2. The device is satisfactorily self-sufficient, because it is based on acceleration and angular change measurements. The system does not propagate radiation and is not drawn to other systems.
3. Navigation air data supplied for whole latitudes, air conditions without ground stations.
4. Position, total velocity, azimuth, vertical velocity, and altitude data are obtained.

Disadvantages of an INS:

1. Position and velocity data corrupt over time. It is valid for both cases of movement or stationary.
2. The device is expensive.
3. It must be tuned first. (First, the position must be entered in the system). So, reference position data are required. Setting is difficult for moving vehicles.
4. The accuracy of the navigation data is related to the vehicle's maneuvers.

Due to the characteristics of accelerometers and gyroscopes, velocity and position data become corrupt over time. Also, system faults increase slowly over time and they are infinite. Unless the fault is improved, an INS fault increases significantly and the system will not be reliable. Therefore, an INS must be used in conjunction with the other navigation systems to limit and damp the faults.

3.3 Inertial Measurement Unit

An INS contains an inertial measurement unit (IMU), including an accelerometer and a gyroscope for all three axes, measuring the linear acceleration and angular velocity of a vehicle with six degrees of freedom (6DOF). By processing the signals from these sensors, it is possible to track the position and orientation of a vehicle.

Measurements of a vehicle's specific forces and rotation rates are the basis of the INS. Measurements can be obtained from triads of gyros and accelerometers that create an IMU. Recent advancements in microelectromechanical systems (MEMS) technology have enabled the production of low-cost inertial sensors. Therefore, the application area of these sensors quickly expanded into the unmanned aerial vehicles (UAVs) industry.

Inertial sensors are nonjammable, nonradiating, and self-consistent, so they cannot be disturbed by any external factors and do not affect anything else surrounding them. However, there can be errors corrupting useful data, even in the highest quality MEMS inertial sensors which are used in UAVs. Regardless of whether the inertial sensor error is caused by internal mechanical imperfections, electronics errors, or other sources, the effect is errors in the indicated outputs of these devices. For the gyros, the major errors are in measuring angular rates. For the accelerometers, the major errors are in measuring acceleration [4]. For both instruments, the largest errors are usually a bias instability [measured in deg/hr for gyro bias drift or micro g (μg) for the accelerometer bias] and scale factor stability [which is usually measured in parts per million (ppm) of the sensed inertial quantity] [5].

3.3.1 Rate Gyros

Rate gyros measure angular velocities, in contrast to free gyros measuring attitude angles. Rate gyros sense the vehicle's angular rate relative to the inertial space [6]. These rate components are the craft angular rate relative to the Earth ω_{nb}, an angular rate as it moves about the spherical Earth ω_{en}, and the angular rate of the Earth itself ω_{ie}. The vector sum of these angular rates ω_{ib} is given by:

$$\omega_{ib}^{b} = \omega_{ie}^{b} + \omega_{en}^{b} + \omega_{nb}^{b} \tag{3.1}$$

There are three types of gyro technology used in today's IMU systems:

- Ring laser gyro (RLG)
- Fiber optic gyro (FOG)
- MEMS

Most current RLG sensors are single DOF sensors requiring three mechanizations for an INS implementation. A single DOF RLG is shown schematically in Fig. 3.1. This figure illustrates a triangular version of the RLG. The gyro includes a laser as a source, a closed-path cavity and mirrors at each intermediate corner in the path, and an interferometer/photodetector. The operation of the gyro is based on optical and electronic phenomena rather than mechanical phenomena [6, 7].

The Sagnac effect can cause a certain rate of rotation inducement with a small difference in the time it takes light to traverse the ring in two directions. Small separation can occur between the frequencies of the counterpropagating beams. Two main patterns can be introduced: standing wave pattern within the ring and beat pattern when those two beams are interfered outside the ring. The net shift of that interference pattern follows the rotation of the unit in the plane of the ring. At very slow rotation rates, RLGs can have the "lock-in" condition. However, they are more accurate than mechanical gyroscopes in most cases. Also, the frequencies of the counterpropagating laser modes get closer and become almost identical if the ring laser is hardly rotating.

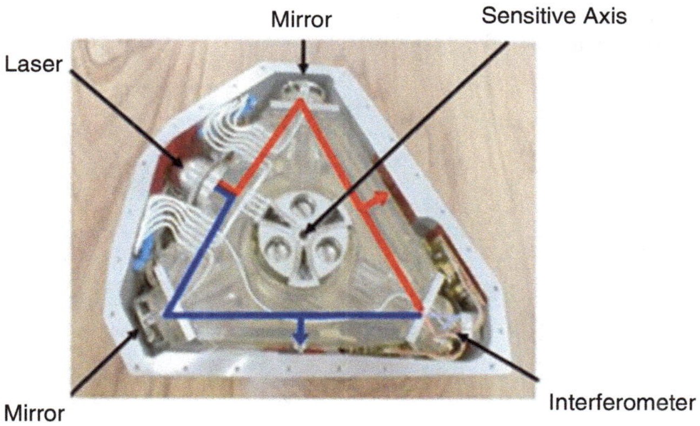

Fig. 3.1 Ring laser gyro

On the other hand, the FOG is a developing gyro technology. FOGs, in comparison to RLGs, require no mechanical dither for their operation and, thus, eliminate a troublesome noise source. They do not require high voltage for the laser plasma, reducing power consumption, and, with the exception of a laser diode for the light source, are composed of passive optical components and, thus, yield extremely high reliability compared to any other currently available technology [6].

As a "rule-of-thumb", an INS equipped with gyros whose bias stability is 0.01 deg/hr will see its navigational error grow at a rate of 1 nmi/hr of operation [8]. Solid-state inertial sensors, such as MEMS devices, have potentially significant cost, size, and weight advantages. The MEMS and interferometry FOG (IFOG) technologies are expected to replace many of the current systems using RLGs and mechanical instruments. The performance of MEMS sensors is continually improving, and they are currently being used in UAVs. The use of MEMS will result in low-cost, high-reliability, small-size, and light-weight inertial sensors being integrated into systems.

3.3.2 Accelerometers

Accelerometers are devices that measure acceleration along the sensing axis. The measurement of acceleration can be integrated into an onboard computer used to derive the UAV's velocity and position. All accelerometers use the principle of sensing the force on a loosely suspended mass, from which the acceleration may be calculated. In comparison to gyros, they have more developed technology. The accelerometer converts acceleration into an electrical signal.

Dynamic acceleration can be measured using an accelerometer. Dynamic acceleration is due to any force except for the gravitational force applied on a rigid body

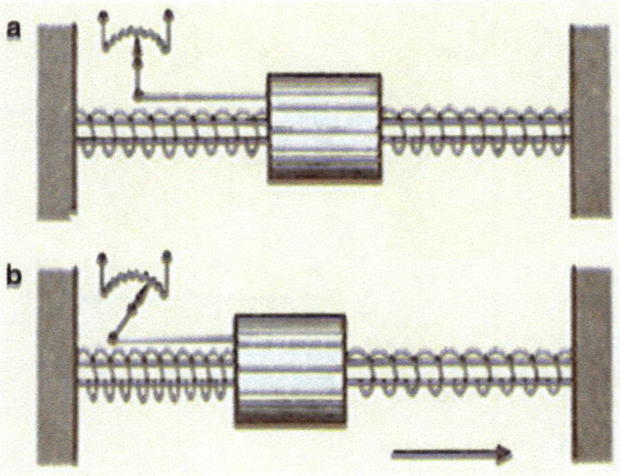

Fig. 3.2 Principle of an accelerometer: (**a**) without acceleration, (**b**) with acceleration

and the static acceleration (or gravitational acceleration) is due to the gravitational force. The output of an accelerometer can be analog or digital. In the analog case, the output voltage is directly proportional to the acceleration. On the other hand, the output of a digital accelerometer can be directly accessed using protocols such as serial interfaces [9]. The principle of an accelerometer is illustrated in Fig. 3.2a, b.

If the body accelerates in the direction indicated in Fig. 3.2, the mass displacement with respect to the body of the device supported on a bar by springs can be recorded. This displacement is proportional to the measured acceleration on the sensing axis. Displacement of the mass generates an electrical signal. For this purpose, potentiometric, self-capacitance converters can be used.

The accelerometer senses the UAV's inertial dynamic acceleration a_i, but is insensitive to the gravitational acceleration. The gravitational field vector g_m, which is the acceleration of mass attraction to the Earth, should be taken into account to determine the total acceleration. Therefore, the acceleration of the UAV can be determined as:

$$a = a_i - g_m \tag{3.2}$$

3.4 Air Data System

All of the air data parameters that are relevant to flight performance are derived by sensing the pressures, temperatures, and flow directions surrounding the vehicle. Free-stream pressures and temperatures are required for the computation of static air temperature, altitude, airspeed, and Mach number [3].

An air data system provides information on quantities such as pressure, altitude, vertical speed, calibrated airspeed, true airspeed, Mach number, static air temperature, and air density ratio [1]. It is, thus, one of the key avionic systems in its own right and forms part of the essential core of avionic subsystems required of all modern aircraft, civil or military. In this section, a short description of the air data system is given.

3.4.1 Air Data Measurements

Air data quantities, such as pressure, altitude, vertical speed, calibrated airspeed, true airspeed, true Mach number, etc., are derived from three basic measurements by sensors connected to probes which measure [1]:

- Total (or pitot) pressure
- Static pressure
- Total (or indicated) air temperature

A basic air data system scheme is presented in Fig. 3.3.

The total pressure P_T is measured by means of an absolute pressure transducer connected to a pitot tube facing the moving airstream. This measures the impact pressure Q_C exerted to bring the moving airstream to rest relative to the pitot tube plus the static pressure P_S of the free airstream, i.e.:

$$P_T = Q_C + P_S. \tag{3.3}$$

The static pressure of the free airstream P_S is measured by an absolute pressure transducer. The transducer is connected to a suitable orifice located where the surface pressure is nearly the same as the pressure of the surrounding atmosphere [1].

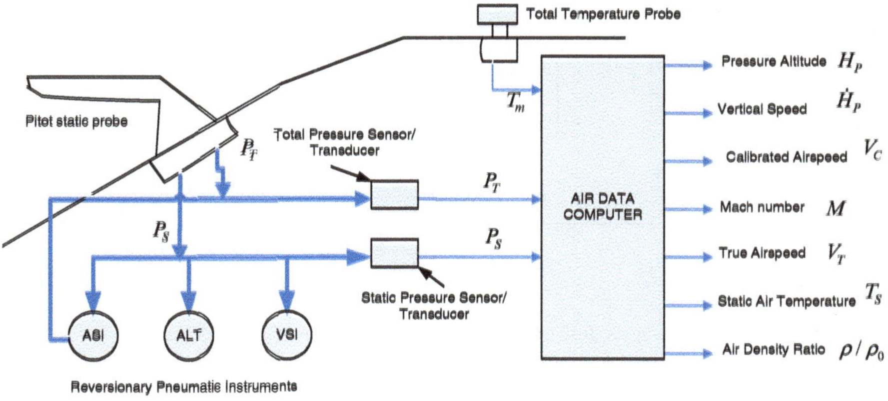

Fig. 3.3 Basic air data system

High-performance military aircraft and UAVs generally have a combined Pitot/static probe which extends out in front of the aircraft, so as to be as far away as practicable from aerodynamic interference effect and shock waves generated by the aircraft structure. Some civil transport aircraft have pitot probes with separate static pressure orifices located in the fuselage, generally somewhere between the nose and the wing. The exact location of the static pressure orifices (and the pitot tubes or probes) is determined by experience and experimentation.

From the measurements of the static pressure and total pressure, it is possible to derive the following quantities [1]:

1. Pressure altitude H_P; this is derived from the static pressure P_S, measured by assuming a "standard atmosphere".
2. Vertical speed H_P; this is basically derived by differentiating P_S.
3. Calibrated airspeed V_C; this is derived directly from the impact pressure Q_C, which is, in turn, derived from the difference between the total and static pressures ($Q_C = P_T - P_S$).
4. Mach number M; this is the ratio of the true airspeed V_T to the local speed of sound A, that is, $M = V_T/A$, and is derived directly from the ratio of the total pressure to the static pressure P_T/P_S. (True airspeed is defined as the speed of the aircraft relative to the air).

A flow diagram of the air data computation procedures, which are explained in detail in [1], is given in Fig. 3.4.

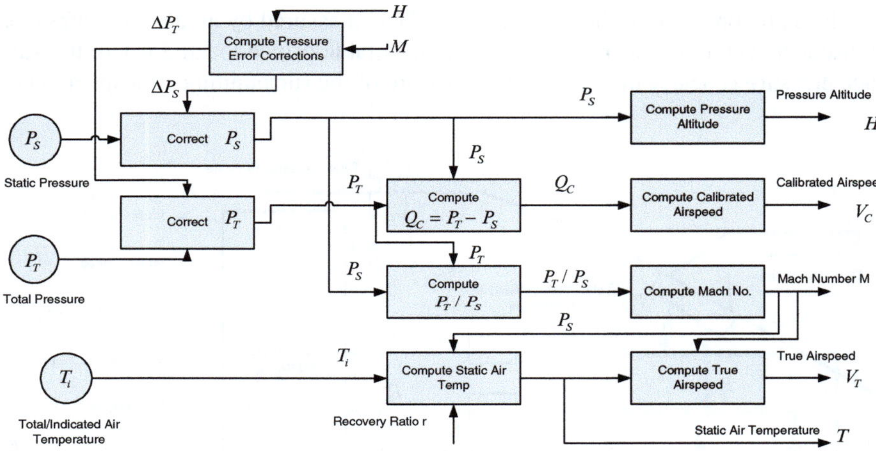

Fig. 3.4 Air data computation flow diagram

3.4.2 Derivation of the True Airspeed Components

The information obtained from the air data system is as follows: true airspeed V_T, angle of attack α, and sideslip angle β. Using these data, the X, Y, and Z components of the true airspeed can be obtained [10]:

$$V_T = \sqrt{V_x^2 + V_y^2 + V_z^2}, \tag{3.4}$$

$$\alpha = \arctan\left(-\frac{V_z}{V_x}\right), \tag{3.5}$$

$$\beta = \arcsin\left(\frac{V_y}{V_T}\right), \tag{3.6}$$

where:

$$\begin{aligned} V_x &= V_T \cos\alpha \cos\beta \\ V_y &= V_T \sin\beta \\ V_z &= -V_T \sin\alpha \cos\beta \end{aligned} \tag{3.7}$$

These speeds are derived according to the aircraft reference system. Later on, these coordinates can be converted to a ground-based reference system for navigational usage.

3.5 Surface Radar

A ground- or surface-based radar station may be used to provide line-of-sight observations of a UAV during flight. These observations usually take the form of measurements of a distance between the UAV and ground radar (range), elevation, and bearing. Figure 3.5 illustrates these navigation parameters geometrically in the Cartesian coordinate system.

The measurements of a UAV's range (D), azimuth (α), and elevation (β) may be expressed in terms of the Cartesian position coordinates of the UAV (x, y, z) as follows:

$$D = \sqrt{x^2 + y^2 + z^2} \tag{3.8}$$

$$\alpha = \tan^{-1}\left\{\frac{y}{x}\right\} \tag{3.9}$$

$$\beta = \tan^{-1}\left\{\frac{z}{\sqrt{x^2 + y^2}}\right\} \tag{3.10}$$

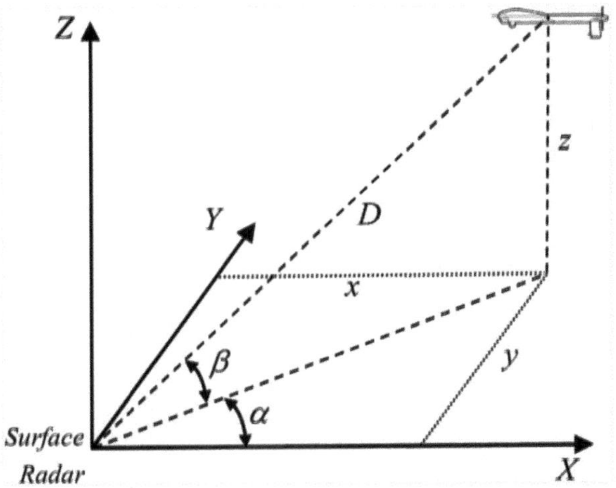

Fig. 3.5 Range, elevation, and bearing observations of an unmanned aerial vehicle (UAV)

Measurement of angle and measurement of range methods are used at the same time in the surface radar [11]. An integrated angle/range measurement method is usually used in radar systems, and it determines the range D of the UAV, azimuth angle α, and elevation angle β. When this method is used, the coordinates of the UAV is determined as the intersection point of the sphere state surface ($D=$ constant), cone state surface (β=constant), and vertical plane suitable for α=constant state surface. UAV coordinates are determined with a single point (ground station) by the help of this method and does not require difficult calculations. The following formulas are used to calculate the UAV's Cartesian coordinates:

$$\begin{aligned} x &= D\cos\beta\cos\alpha \\ y &= D\cos\beta\sin\alpha \\ z &= D\sin\beta \end{aligned} \tag{3.11}$$

Range, azimuth, and elevation angles are determined by radar measurements. A single radar station is sufficient to use this method.

Advantages of the angle/range integrated method:

- A single station is needed to measure the coordinates
- Coordinates are calculated as simple
- The method provides the required accuracy in whole measurement intervals
- Data processing is easy in the system

Disadvantage of the angle/range integrated method:

- The accuracy of the UAV coordinate measurement decreases with distance away from the surface radar

3.6 Altitude Measurements

3.6.1 Types of Flight Altitudes

An altimeter is a device that measures the flight altitude of aircraft on the Earth. Three types of altitudes are considered during flight:

- Absolute altitude = flight altitude according to sea level
- Relative altitude = flight altitude with respect to the takeoff or landing site
- Real altitude – altitude over the flight area

Descriptions of the altitudes are given in Fig. 3.6.

The value of absolute altitude is required for setting corridors to flight routes and flight tests of aircrafts and engines.

The relative altitude must be known for taking off and landing.

The real altitude is for arbitrary flight status.

3.6.2 Radio Altimeter

The measurement of flight altitude via radiotechnical methods depends on measuring the time interval elapsed during the signal's path from the aircraft to the Earth [3]. Two antennas are installed in the aircraft in order to use a radiotechnical method – transmitter A_1 and receiver A_2. Antennas are often placed on the wings of aircraft (Fig. 3.7a).

Transmitter antenna A_1 sends out radio waves, which are reflected when they reach the Earth's surface. Some of the reflected waves are received by the aircraft's receiver antenna, A_2. The time duration between the signal's transmission to Earth and its return is determined as presented below:

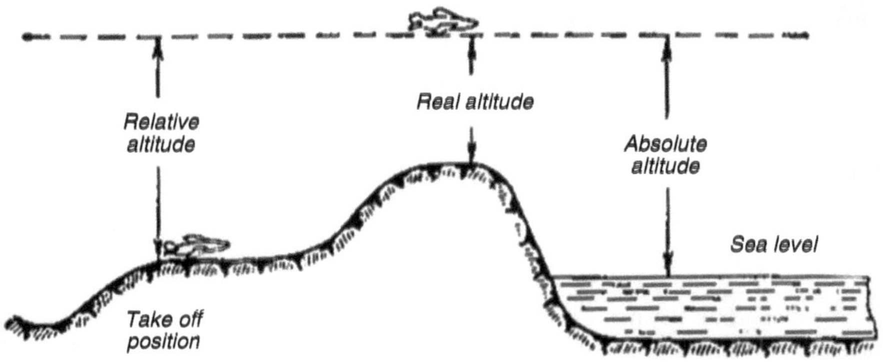

Fig. 3.6 Flight altitude types

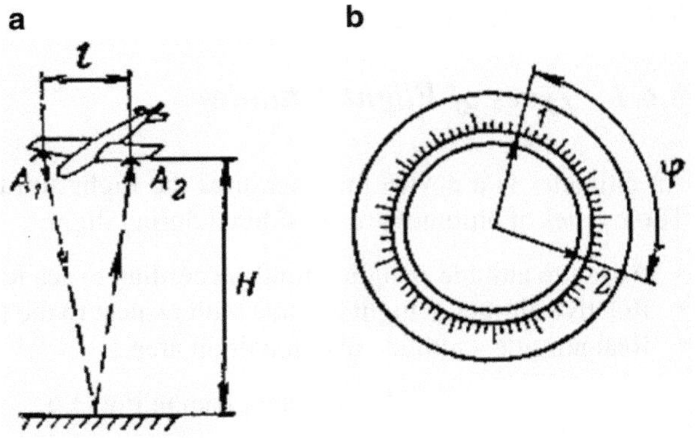

Fig. 3.7 Radio altimeter: A_1, A_2 – transmitter and receiver antennas

$$t_1 = \frac{2H}{c} \tag{3.12}$$

where H is the flight altitude and $c = 3\ 10^8$ m/s is the propagation velocity of radio signals.

The registration of two signals' receiving times is used to measure the propagation time of radio signals: the signal from the transmitter and the reflected signal. The first signal reaches the receiver antenna after time $t_2 = l/c$ and the second after time $t_1 = 2H/c$. At that time, the real altitude of the flight will be equal to:

$$H = \frac{(c\tau + l)}{2}, \tag{3.13}$$

where $\tau = t_1 - t_2$ and l is the range between the transmitter and receiver antennas.

Expression (3.13) is used to steady radio altimeters or radio altimeters with impulse teleportation.

When high altitudes are measured ($H \geq 1,500$ m), the power of the transmitted signal decreases significantly, so, for such altitudes, impulse teleportation must be run after steady teleportation.

An altimeter's receiver takes the reflected impulses from the aircraft and the Earth and, after strengthening the impulses, they are passed to an electron gun (see Fig. 3.7b). A circular movement with angular velocity ω is acted upon the beam to find the first impulse towards the screen and the reflected second impulse. The angle between the impulses φ (Fig. 3.7b) is calculated via the following formula:

$$\varphi = \omega\tau = \frac{2H\omega}{c} \tag{3.14}$$

For increasing the measurement accuracy, the angular velocity ω of the beam must be increased. In this case, the height of the device's whole pointer must be appropriate for the measurement interval.

3.6.3 Barometric Altimeter

In airborne vehicles flying at low altitudes, it is very important to provide accurate altitude information continuously. For this purpose, barometric (baro) altimeters are commonly used nowadays. They can measure the atmospheric pressure value, according to the relationship between the height and the air data, such as pressure and temperature, and indirectly calculate the height. As is known [12], the baro altimeter is an excellent sensor for vertical height measurement during straight-and-level flight segments and provides short-term accurate vertical altitude information. In general, barometers are initialized using the height, temperature, and pressure at the starting point of the vehicle in which the barometer is installed. However, as time elapses and as the aircraft moves to another place, the characteristics of the atmosphere around the vehicle are changed, which produces large barometric errors [13]. To surmount this problem, a baro altimeter must be aided by other altitude sensors, such as GPS, radio altimeter, etc.

Altitude is determined via the barometric altitude measurement equation. Assuming that air is an ideal gas and that the gradient of temperature as a function of altitude (i.e., the atmospheric lapse rate) is known, then the altitude can be computed via the following expression [14]:

$$h^B = \frac{T_0}{T_{\text{grad}}}\left[1 - \left(\frac{p}{p_0}\right)^{\frac{T_{\text{grad}}R}{g}}\right] \tag{3.15}$$

where h^B is the barometric altitude, T_0 is the temperature at sea level, T_{grad} is the lapse rate, p_0 is the pressure at sea level, p is the pressure measured at altitude h^B, R is the universal gas constant, and g is the local gravity.

The variation of atmospheric pressure with height is given in Fig. 3.8.

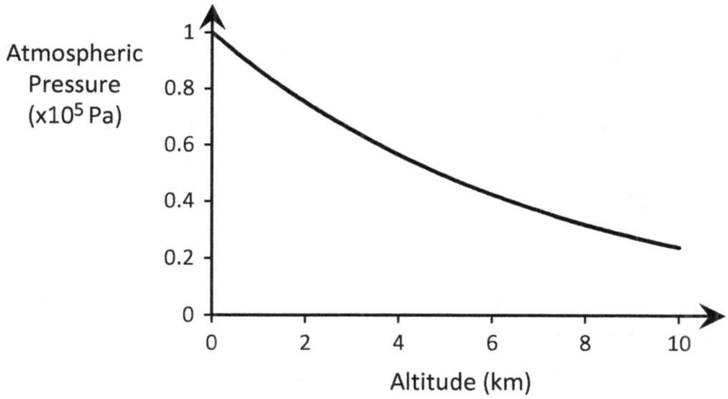

Fig. 3.8 Variation of atmospheric pressure with altitude

The conversion of the measured air pressure to altitude given in Eq. (3.15) is based on a theoretical standard atmosphere and the assumption that air is an ideal gas. More precisely, the standard atmosphere is defined as follows [15]:

- Air is an ideal gas with gas constant $R = 287\ \mathrm{J}/(\mathrm{Kg} \cdot \mathrm{K})$
- The pressure at sea level is $p_0 = 29.92$ in. – Hg
- The air temperature at sea level is $T_0 = 15\ °\mathrm{C}$
- The temperature gradient (lapse rate) is $T_{grad} = 0.0065\ °\mathrm{C/m}$

The pressure at sea level p_0 is not constant but varies from day to day and location to location. In practice, therefore, a value of p_0 which makes the altitude estimated using (1) correct is input into the altimeter. This value of p_0 is known as the altimeter setting.

The main source of error in barometric altitude measurement arises from differences between the true and modeled atmospheric temperature and pressure. For standalone barometry, altitude errors can be several hundred meters. For differential barometry, the error increases with distance from the reference station and age of the calibration data [16].

3.7 Speed-over-ground and Drift Angle Measurements via the Doppler Method

Doppler radar provides speed-over-ground measurement of aircraft. In addition, the aircraft drift angle between the velocity vector W and its component in the x direction W_x can be measured by the Doppler method. A Doppler radar operates by transmitting a narrow beam of microwave energy to the ground and measures the frequency shift that occurs in the reflected signal as a result of the relative motion between the aircraft and the ground [16].

Speed-over-ground (W) and drift angle (β) can be measured by the use of the Doppler method. The method depends on the transmission of definite frequency signals and receiving the reflected signals with altered frequency. The change infrequency is related to the flight velocity. The frequency change caused by the Doppler effect is measured by a device calibrated by velocity units.

Assume that transmitter and receiver systems exist in the aircraft with W total velocity. Radio signals with frequency f sent by the transmitter arrive at the Earth's surface and are then reflected and picked up by the aircraft.

The frequency of the signals observed by the aircraft's receiver is $f_1 = f + 2F_D$, where $F_D = (W/\lambda)\cos\gamma$ is the Doppler frequency drift, λ is the wavelength, and γ is the inclination angle of the beam.

The error of the method is increased by inaccurate filtration of the signal and the roll angle of the aircraft. For instance, when $\gamma = 70°$ and the roll angle is 1°, the error of the velocity measurement is 4.7 %. Compensation methods presented below are used to reduce the error [11]:

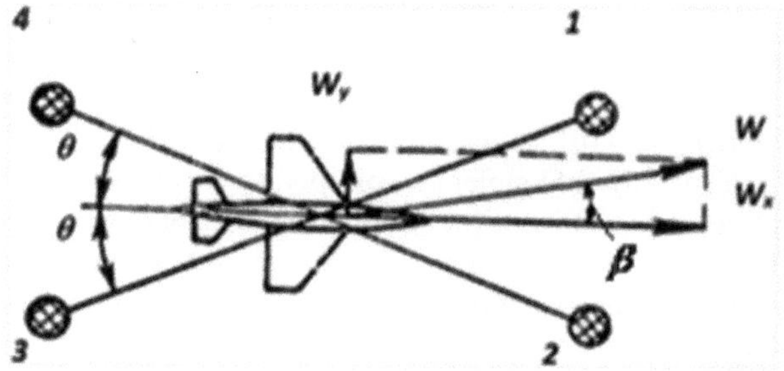

Fig. 3.9 Doppler velocity measurement: scheme of a measurement system with four beams

1. Stabilization of antennas according to the Earth's vertical axis
2. Consideration of corrections according to rolling analytically
3. Application of multibeam systems

The first two methods do not provide measurements of high accuracy. When the third method is used in case of the determination of two beams symmetrically according to the vertical, it is possible that the measurement error of the velocity is decreased to 0.1 % in 1° rounding.

In a Doppler velocity and drift angle measurement device, a four-beam antenna is used (Fig. 3.9).

In such a system, four antennas send radio signals to areas 1, 2, 3, and 4 on Earth. Therefore, the sum of Doppler frequencies of two pairs of channels, (1–3) and (2–4), are, respectively:

$$\begin{aligned} f_{d(1+3)} &= 4\left(^{W}/_{\lambda}\right)\cos\gamma\cos(\theta+\beta) \\ f_{d(2+4)} &= 4\left(^{W}/_{\lambda}\right)\cos\gamma\cos(\theta-\beta) \end{aligned} \quad (3.16)$$

By calculating the sum and subtracting the divided frequencies, we can write:

$$\begin{aligned} f_{d(1+3)} + f_{d(2+4)} &= k_1 W\cos\beta \\ f_{d(1+3)} - f_{d(2+4)} &= k_2 W\sin\beta \end{aligned} \quad (3.17)$$

Here, $k_1 = (8/\lambda)\cos\gamma\cos\theta$ and $k_2 = (8/\lambda)\cos\gamma\sin\theta$.

The total flight velocity W and drift angle β are determined via Eq. (3.17).

When the antenna system is rotated by a certain drift angle, the bisector of 2θ angle of the system will be in the direction of the total flight velocity vector, so the drift angle will be equal to zero. It is clear that the equations:

$$\begin{aligned} f_{d(1+3)} + f_{d(2+4)} &= k_1 W \\ f_{d(1+3)} - f_{d(2+4)} &= 0 \end{aligned} \quad (3.18)$$

are valid.

The second equation is the difference signal. This signal is used in schemes which determine W and β automatically.

The standard deviations of Doppler measuring devices is 0.2 % in respect of the velocity and 0.1 % in respect of the drift angle.

Typically, a Doppler sensor operating over land is able to provide measurements of velocity to an accuracy of about 0.25 % or better. Performance is degraded during flights over water owing to poor reflectivity, scattering of the reflected signal, giving rise to a bias in the measurement, wave motion, tidal motion, and water currents. However, this navigation aid offers good long-term stability and a chance to bind the position and velocity estimates provided by an inertial navigation system [16].

3.8 Magnetic Measurements

The variations in the Earth's geomagnetic field can be internal, which are related to the dynamo within the Earth's core and crust. Inner dynamo variations, which are long-term variations within the core, are called secular variations. On the other hand, solar wind–geomagnetic dipole interaction variations are relatively short-duration variations on the order of months, days, and hours. The Earth has a magnetic field similar to that of a bar magnet, with poles located close to its axis of rotation. This means that the direction of the horizontal component of the Earth's magnetic field lies close to true north, and the magnetic north, as determined by a magnetic field sensor, or compass, can be used as a working reference [16]. The angle between true north and the magnetic north is not constant. It varies with the position of the observation on the Earth and slowly with time, although it is possible to compensate for this. The direction of the Earth's magnetic field at any point on its surface is defined in terms of its orientation with respect to true north, known as the angle of "magnetic declination", and its angle with respect to the horizontal is called the angle of "dip". These angles are illustrated in Fig. 3.5. Magnetometers are widely used in aerospace as attitude determination devices. These sensors provide the direction of the Earth's magnetic field and its magnitude; therefore, they are useful for determining the attitude of flying vehicles. Moreover, their low mass and low power consumption make them attractive for UAVs. For these reasons, most UAVs have magnetometers as part of their basic sensor package.

In the last decade, the popularity of UAVs, due to their low cost and weight, has increased significantly. That has brought about a search for lighter but more accurate sensors. Under these circumstances, the three-axis magnetometer (TAM) has become very attractive because of its advantages, such as providing continuously available two-axis attitude measurements, relative low cost, and almost insignificant power demand (Fig. 3.10).

In the absence of local magnetic disturbances, magnetometers sense the components of the Earth's magnetic field (H_0) acting along its sensitive axis. A TAM may be mounted in a UAV to sense the components of the Earth's magnetic field along its principal axes, (H_x, H_y, H_z). Expressing a UAV's attitude, with respect to the local

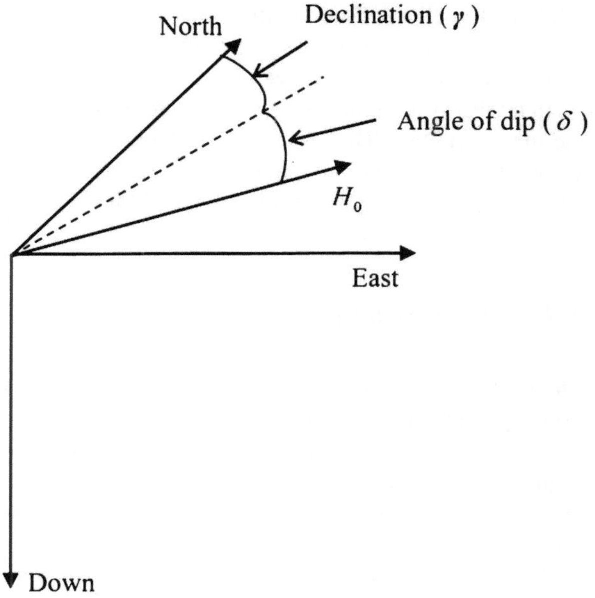

Fig. 3.10 Angle of "magnetic declination" and angle of "dip"

coordinate frame, can be done as a direction cosine matrix, C_n^b. The relationship between the magnetic measurements and the attitude of the UAV may be written as:

$$\begin{bmatrix} H_x \\ H_y \\ H_z \end{bmatrix} = C_n^b \begin{bmatrix} H_0 \cos\delta \cos\gamma \\ H_0 \cos\delta \sin\gamma \\ H_0 \sin\delta \end{bmatrix}, \tag{3.19}$$

where δ is the angle of dip and γ is the angle of declination. If the angle of dip is known, as well as the variation between true and magnetic norths, estimates of a UAV's attitude can be made from the magnetometer measurements.

The aircraft in which the magnetic sensor is mounted will almost always have a magnetic field that cannot be distinguished from that of the Earth's magnetic field and, consequently, it is necessary to compensate for this effect as well. Therefore, in order to estimate the components of the geomagnetic field vector for determining the attitude of a UAV accurately, the magnetometer should be calibrated precisely.

3.9 Satellite Radio Navigation

The Global Positioning System (GPS) is a space-based global navigation satellite system developed by the USA. The GPS provides reliable location and time information in all weather conditions, at all times, and anywhere. The system uses a total of 24 satellites. Four satellites are necessary to calculate the position of the user. There are also other similar systems used or planned to be used, such as GLONASS (Russia) and Galileo (European Union).

In a GPS system, the GPS receiver chooses the best available satellites to obtain the measurements. The system is available as two services, a standard positioning system (SPS) for civilians and a precise positioning service (PPS) for military users. The order of accuracy of SPS is 10 m and that of PPS is around 3 m.

3.9.1 GPS Structure

The GPS consists of three parts: the space segment, the control segment, and the user segment [3]. The structure of the GPS is shown in Fig. 3.11.

The U.S. Air Force develops, maintains, and operates the space and control segments. GPS satellites broadcast signals from space, which each GPS receiver uses to calculate its three-dimensional location (latitude, longitude, and altitude) plus the current time.

The space segment is composed of 24–32 satellites in medium Earth orbit and also includes the boosters required to launch them into orbit. The control segment is composed of a master control station, ground antennas, and monitor stations. The user segment is, in our case, composed of UAVs.

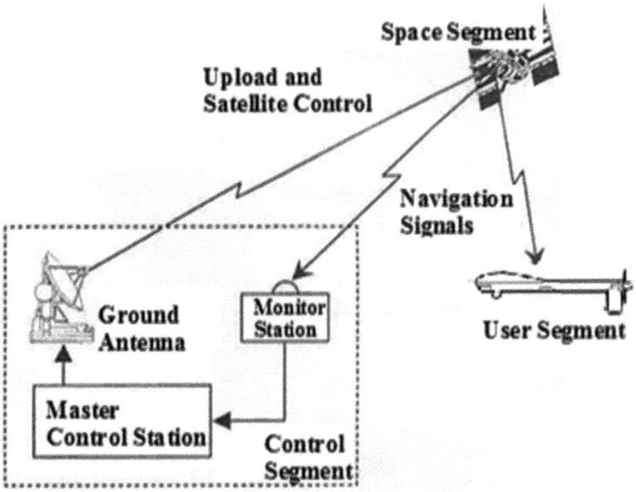

Fig. 3.11 Satellite radio navigation system structure

3.9.2 *Basic Concept of GPS*

A GPS receiver calculates its position by precisely timing the signals sent by the GPS satellites. Each GPS satellite continually transmits messages, which include:

- The time the message was transmitted
- Precise orbital information (the ephemeris) of satellites
- The general system health

The receiver utilizes the messages it receives to determine the transit time of each message and computes the distances to each satellite. These distances, along with the satellites' locations, are used to compute the position of the receiver. This position is determined as the geographical coordinate system latitude, longitude, and elevation. Many GPS units give derived information, such as the direction and speed of the user, calculated from position changes.

A very small clock error, which is the discrete time scale of the user and the satellite system multiplied by the very large speed of light—the speed at which satellite signals propagate—results in a large positional error (clock bias). Therefore, receivers use four or more satellites to solve for the four-parameter determination: clock bias and receiver's location as latitude, longitude, and elevation.

Today, satellite navigation systems are used for solving the navigation problems of most of UAVs. In this technique, the position of the object is generally detected with the help of distance measuring methods (distances from object to $n \geq 4$ satellites are measured) (see Fig. 3.12).

Then, the coordinates of the object will be on the intersection of the sphere surfaces with radiuses equal to the measured distances; see [3, 11, 17]. The equation system that consists of $n \geq 4$ nonlinear equations, which are obtained by this approach, can be solved using a computer with the aid of iteration methods. However, the distance equations used for formerly stated purposes have a term other than the unknown coordinates of the user, which results from the discrete

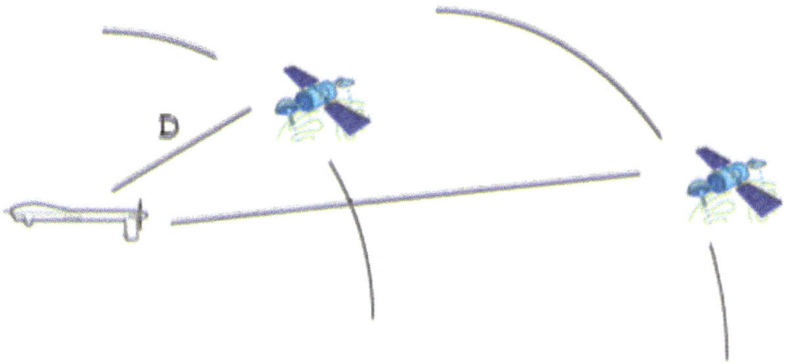

Fig. 3.12 Scheme of locating an object using distance measuring

time scale of the user and the satellite system [3, 18]. In this case, this term and the user coordinates are determined as the intersection point of the n different sphere position surfaces (R = constant). In a simple case, the user coordinates and clock bias can be obtained by solving the four nonlinear algebraic equations below:

$$(x_i - x)^2 + (y_i - y)^2 + (z_i - z)^2 = (D_i - \delta t)^2 \quad i = \overline{1,4} \tag{3.20}$$

where x_i, y_i, and z_i are the Descartes coordinates of the ith satellite, in the geocentric axis system, 4 is the minimum number of satellites, x, y, and z are the Descartes coordinates of the object (users) in the geocentric coordinate system, and δt is the distance measuring errors, resulting from the time differences between the user and the satellite (clock bias).

UAVs are now able to work in different environments easily using a GPS receiver. This allows the use of MALE UAVs (a medium-altitude, long-endurance UAV) and HALE UAVs (a high-altitude, long-endurance UAV) in every part of the world. However, in some places, by jamming or with local conditions, GPS receivers may receive weak signals. Thus, depending solely on GPS may cause problems in the navigation of UAVs. To overcome this problem, dead reckoning systems can be used. A filter, such as a Kalman filter, may be used to integrate the measurements of GPS with dead reckoning devices to obtain better measurements.

3.10 Vision-based Systems

Visual sensing can also provide data for position estimation. These kinds of systems have the advantage of using visual sensors that actually sense the real physical data to comprehend the results in an easier way. The data from the visual sensors can also be combined with the data from INS, GPS, and magnetometers in order to supply a better understanding to the operator. Improved digital imaging techniques make the use of visual sensors in small UAVs navigation more practical and simpler. Also, using methods for emulating flying insects is a new approach of developed systems in visual sensing. One method is the use of a newly developed artificial compound insect eye, which is a system that replicates the eye structure of an insect. This system uses many small-sized lenses with hemispherical shapes. The camera chips in this system are made of hard and soft composite materials, which allow stretching and bending, therefore acting like an insect's eye, as desired [19–21].

3.11 Simultaneous Localization and Mapping (SLAM)

SLAM algorithms are used to develop systems that are terrain aided. These are used for map building. Using generated maps, these systems help other sensors. A few of these systems are used on robotic systems that depend on visual sensors. Visual motion estimation techniques that use stereovision and visual features tracking have been proposed in the context of ground rovers [22]. Stereovision and monocular systems are also used in different studies [23]. These types of systems allow UAVs to operate in all environments, including those in which they face GPS issues.

3.12 Measurement Fault Classification and Fault Modeling

A "fault" is to be understood as an unexpected change of any system function, although it may not represent physical failure. Such a fault or malfunction hampers or disturbs the normal operation of the system, thus causing an unacceptable deterioration in performance or even leading to dangerous situations [23].

"A fault is an unpermitted deviation of at least one characteristic property (feature) of the system from the acceptable, usual, standard condition" [24].

Based on this definition, a fault corresponds to an abnormal behavior of the system, which may not affect its overall functioning, but may eventually lead to a failure.

A fault may be small or hidden, and, therefore, difficult to detect, isolate, and identify.

"A failure is a permanent interruption of a system's ability to perform a required function under specified operating conditions" [24].

Resulting from one or more faults, a failure of the device is, therefore, an event that terminates the functioning of a device.

In this book, we use the term "fault" rather than "failure" to denote a malfunction rather than a catastrophe. The concept of failure is known as complete breakdown of a system component or function, whilst the concept of fault may be used to indicate that a malfunction may be tolerable at its current step. A fault must be diagnosed as early as possible to prevent any serious consequences.

A sensor fault occurs as soon as the measurement data deviate from the physically measured process by more than the noise uncertainty. Bias, noise increment, or wrong scaling factors are also classified as a sensor fault. The possible sensor faults are given in Fig. 3.13.

The mathematical representation of the above sensor faults is as follows [25]:

$$y_i(t) = \begin{cases} x_i(t), \forall t \geq t_0 - \text{No. failure} \\ x_i(t) + b_i, \dot{b}_i(t) = 0, b_i(t_{Fi}) \neq 0 - \text{Bias} \\ x_i(t) + b_i(t), |b_i(t)| = c_i t, 0 < c_i << 1, \forall t \geq t_{Fi} - \text{Drift} \\ x_i(t) + b_i(t), |b_i(t)| \leq \overline{b}_i, \dot{b}_i(t) \in L^{\infty}, \forall t \geq t_{Fi} - \text{Loss. of. accuracy} \\ x_i(t_{Fi}), \forall t \geq t_{Fi} - \text{Sensor. freezing} \\ k_i(t) x_i, 0 < \overline{k} \leq k_i(t) \leq 1, \forall t \geq t_{Fi} - \text{Calibration. error} \end{cases}$$

Here, t_{Fi} is the time-off fault occurrence on the ith sensor and b_i is its accuracy coefficient, such that $b_i \in \left[-\overline{b}_i, \overline{b}_i\right]$, where $\overline{b}_i$>0. Furthermore, $k_i \in \left[\overline{k}_i, 1\right]$, where $\overline{k}_i > 0$, denotes the minimum sensor effectiveness, as can be seen. Except the freezing case, the following general mathematical model can be represented:

$$y = K_m x + B, \tag{3.21}$$

where K_m is a positive-definite diagonal matrix whose elements are slowly varying within $\left[\overline{k}_i, 1\right]$ and elements of vector B slowly vary within $\left[-\overline{b}_i, \overline{b}_i\right]$ [26].

If the parameter is observed incase of measurement noise increment, then the measurement equation can be presented in the form:

$$y_i(t) = x_i(t) + \sigma_i(t)v_i(t) \tag{3.22}$$

where $v_i(t)$ is the random measurement noise and $\sigma_i(t) > 1$ characterizes the measurement noise increment. This type of sensor fault is presented in Fig. 3.13f. The measurement noise corresponding to normal operation of the sensor in the figure is shown as the red line, and the measurement noise increment is shown by the black line.

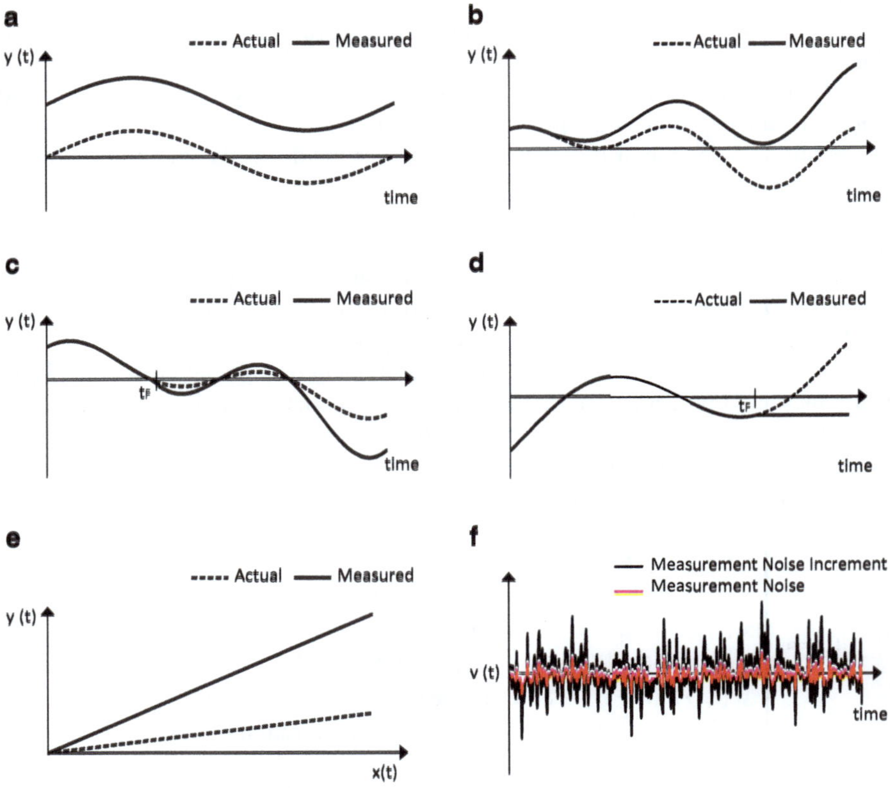

Fig. 3.13 Several types of sensor faults: (**a**) sensor bias, (**b**) sensor drift, (**c**) loss of accuracy, (**d**) frozen sensor (occurring after t_F), (**e**) calibration error, and (**f**) noise increment

For example, all types of accelerometers and gyros exhibit biases, scale factor and misalignment errors, and random noise to a certain extent [27]. The bias is a constant error which manifests in the measurement results. Mostly, accelerometers and gyros have bias errors. It is independent of the underlying specific force and angular rate.

The scale factor error is the departure of the input–output gradient of the instrument from unity following unit conversion by the IMU. The accelerometer output error due to the scale factor error is proportional to the true specific force along the sensitive axes, while the gyro output error due to the scale factor error is proportional to the true angular rate about the sensitive axis [27].

Misalignment errors in all types of IMUs arise from the misalignment of the sensitive axes of the inertial sensors with respect to the body frame orthogonal axes because of manufacturing limitations.

To give an example regarding this subject, the reasons for INS sensor faults may be listed as follows:

- Gyroscope drift faults caused by temperature changes, acceleration, magnetic field, and vibration
- Gyroscope measurement faults with nonlinearity and asymmetry
- Accelerometer bias because of temperature changes and vibrations
- Accelerometer measurement faults with nonlinearity and asymmetry
- The input axes of gyroscopes and accelerometers are not perpendicular to each other
- Initial state faults (must be considered that position and velocity errors are proportional to time) etc.

Sensor faults may cause substantial performance degradation of a decision-making system. For example, in a UAV flight control system, sensors are used to measure directly the UAV states. Thus, the presence of faults in sensors may deteriorate state estimates and, consequently, cause inaccurate flight control.

A fault monitoring system which is used to detect faults and diagnose their location and significance within a system is called a "fault diagnosis system". Such a system normally consists of the following tasks [28]:

- Fault detection: to make a binary decision – either that something has gone wrong or that everything is fine
- Fault isolation: to determine the location of the fault, e.g., which sensor or actuator has become faulty
- Fault identification: to estimate the size and type or nature of the fault

A fault-tolerant flight control system is capable of controlling UAV dynamics with satisfactory performance, even if there is one or more faults.

The fault-tolerant state estimation system of UAVs in this book is capable of detecting and diagnosing faults in the UAV control system and adequately compensate for failures.

References

1. Collinson RPG (1996) Introduction to avionics. Chapman & Hall, London
2. Lin C-F (1991) Modern navigation, guidance, and control processing. Prentice Hall, Englewood Cliffs
3. Kayton M, Fried WR (1997) Avionics navigation systems. Wiley, New York
4. Prestero T (2001) Verification of a six-degree of freedom simulation model for the REMUS autonomous underwater vehicle. M.Sc. thesis in Ocean and Mechanical Engineering, MIT, Cambridge, MA
5. Stutters L, Liu H, Tiltman C, Brown DJ (2008) Navigation technologies for autonomous underwater vehicles. IEEE T Syst Man Cybern Part C Appl Rev 38(4):581–589
6. Titterton DH, Weston JL (2004) Strapdown inertial navigation technology, vol 207, Progress in astronautics and aeronautics. MIT Lincoln Laboratory, Lexington
7. Volk C, Lincoln J, Tazartes D (2008) Northrop Grumman's family of fiber-optic based inertial navigation systems, USA. http://www.nsd.es.northropgrumman.com. Date retrieved 19 Jan 2012
8. Schmidt GT (2007) INS/GPS technology trends. NATO SET-104 symposium, Antalya
9. Bekir E (2007) Introduction to modern navigation systems. World Scientific, Hackensack
10. Vasil'chenko KK et al (1996) Flight tests on aircraft. Mashinostroenie, Moscow (in Russian)
11. Hajiyev C (1999) Radio navigation. Istanbul Technical University, Istanbul (in Turkish)
12. Tang W, Howell G, Tsai Y-H (2005) Barometric Altimeter short-term accuracy analysis. IEEE Aerospace and Electronic Systems Magazine, December, pp 24–26
13. Whang I-H, Ra W-S (2007) Barometer error identification filter design using sigma point hypotheses. In: Proceedings of the international conference on control, automation and Systems (ICCAS) 2007, Seoul, Korea, October 17–20 2007, pp 1410–1415
14. Jan S-S, Gebre-Egziabher D, Walter T, Enge P (2008) Improving GPS-based landing system performance using an empirical barometric altimeter confidence bound. IEEE T Aero Elec Sys 44(1):127–146
15. Von Mises R (1959) Theory of flight. Dover, New York
16. Groves PD (2008) Principles of GNSS, inertial, and multisensor integrated navigation systems. Artech House, Boston
17. Parkinson BW, Spilker JJ Jr (eds) (1996) Global positioning system: theory and applications, vol II. AIAA, Inc., Washington, DC
18. Zarchan P, Musoff H (2000) Fundamentals of Kalman filtering: a practical approach. AIAA, Inc., Washington, DC
19. Bento MF (2008) Unmanned air vehicles: an overview. Inside GNSS, Jan–Feb 2008
20. http://spectrum.ieee.org/robotics/robotics-hardware/insecteye-camera-offers-wideangle-vision-for-tiny-drones, Jeremy Hsu, posted 01.05.2013 – accessed on 3 Dec 2014
21. Conta G, Doherty P (2009) Vision-based unmanned aerial vehicle navigation using georeferenced information. EURASIP J Adv Signal Process vol 2009, article ID 387308, Hindawi Publishing Corporation, Cairo, Egypt
22. Mallet A, Lacroix S, Gallo L (2000) Position estimation in outdoor environments using pixel tracking and stereovision. In: Proceedings of the IEEE international conference on robotics and automation (ICRA), San Francisco, CA, April 2000, pp 3519–3524
23. Lemaire T, Berger C, Jung I-K, Lacroix S (2007) Vision-based SLAM: stereo and monocular approaches. Int J Comput Vis 74(3):343–364
24. Isermann R (2006) Fault-diagnosis systems: an introduction from fault detection to fault tolerance. Springer, Berlin/Heidelberg
25. Boskovic JD, Mehra RK (2002) Stable adaptive multiple model-based control design for accommodation of sensor failures. In: Proceedings of the 2002 American control conference, anchorage, AK, May 2002, pp 2046–2051

26. Sobhani-Tehrani E, Khorasani K (2009) Fault diagnosis of nonlinear systems using a hybrid approach, vol 383, Lecture notes in control and information sciences. Springer Science + Business Media, LLC, Dordrecht
27. Groves PD (2013) Principles of GNSS, inertial, and multisensor integrated navigation systems, 2nd edn. Artech House, Boston/London
28. Hajiyev C, Caliskan F (2003) Fault diagnosis and reconfiguration in flight control systems. Kluwer Academic Publishers, Boston

References

Chapter 4
Estimation of Unmanned Aerial Vehicle Dynamics

4.1 Introduction

In this chapter, we will consider the solution to a class of linear minimum-error-variance sequential state estimation algorithms originally developed by Kalman in a classical work [1]. The Kalman filter, also known as linear quadratic estimation (LQE), is an algorithm that uses a series of measurements observed over time, containing random noise and other inaccuracies, and produces estimates of unknown variables that tend to be more precise than those based on a single measurement alone. Mathematically, the Kalman filter is a system of first-order ordinary differential equations with quadratic nonlinearities, which are solved on digital computers. Kalman filter algorithms, or one of their many extensions and variations, have been applied in numerous practical situations, including navigation, space guidance, motion control, and orbit determination. The Kalman filter is used for the following tasks [2]:

1. Filtration (finding a better value of a measured parameter)
2. State estimation (calculating the state vector of an aircraft)
3. System identification (calculating the unknown parameters of the mathematical model of an aircraft)
4. Prediction (predicting the state vector of an aircraft)
5. Smoothing (estimators that take into account both the past and the future)
6. Sensor fusion (integrating more than one source of information)
7. Fault detection and diagnosis (detection, isolation, and identification of the faults that occur in an aircraft's systems)

We shall begin our treatment of the Kalman filter with the discrete-time version of the problem; that is, observation of a discrete dynamic system. For simple problems, the discrete algorithms can be manipulated by hand and considerable insight may be gained. The step-by-step processing of information lends itself to a simple development.

C. Hajiyev et al., *State Estimation and Control for Low-cost Unmanned Aerial Vehicles*, DOI 10.1007/978-3-319-16417-5_4

The equations of motion for an unmanned aerial vehicle (UAV) are given in Chap. 2. In this chapter, the equations of motion are further examined and their linearization is discussed in detail for the stability analysis. In this procedure, the first step is investigating the trim conditions.

4.2 The Optimal Linear Discrete Kalman Filter

In this chapter, we introduce and briefly discuss the mathematical model and the relations of the linear discrete-time Kalman filter and its application to UAV dynamics.

Let us take a linear discrete dynamic system. The system's dynamic state equation defines the dynamics of the system, whereas the measurement equation defines the generation mechanism of the measurement. The equations for a linear system are written as follows:

State equation:

$$x_{k+1} = F_{k+1}x_k + B_{k+1}u_k + w_k \tag{4.1}$$

Measurement equation:

$$z_k = H_k x_k + v_k \tag{4.2}$$

where x_k is the n-dimensional state vector of the system, F_{k+1} is the $n \times n$ transfer matrix of the system, u_k is the p-dimensional vector of the deterministic control input, B_{k+1} is the $n \times p$ control distribution matrix, w_k is the n-dimensional random Gaussian noise vector (system noise) with zero mean and correlation matrix $E\left[w_k w_j^{\mathrm{T}}\right] = Q_k \delta_{kj}$ is the statistic average operator, δ_{kj} is the Kronecker symbol:

$$\delta_{kj} = \begin{cases} 1, k = j, \\ 0, k \neq j. \end{cases}$$

z_k is the s-dimensional measurement vector, H_k is the $s \times n$-dimensional measurement matrix of the system, and v_k is the s-dimensional measurement noise vector with zero mean and correlation matrix $E\left[v_k v_j^{\mathrm{T}}\right] = R_k \delta_{kj}$.

The mean of the initial condition x_0 is $\overline{x}_0$ and the correlation matrix is P_0. There is no correlation between the system noise w_k and the measurement noise v_k: $E\left[w_k v_j^T\right] = 0, \forall k, j$.

It is desired to find the value of the state vector according to the sequence of the y_k measurements vectors. Linear filters theory based on the Kalman filter must be used for that purpose.

Depending on the application, one might want to obtain an estimate of the state at a certain time. If the state is estimated for some time in the future, the process is called prediction. If the estimate is made using all measurements up to and including the current moment, it is called filtering. If an estimate is made for some time in the past using measurements up to the current moment, the process is called smoothing.

In this chapter, we limit ourselves to prediction and filtering.

4.2.1 *Optimal Kalman Filter (OKF) Equations*

Generally, the filtering process of Kalman filters can be examined in two distinct phases and if the equations are presented in the following way [3]:

Time Update (Prediction) The phase where the estimations of the preceding step is used for producing the estimations of the present step. It can be thought of as preparation to the estimation.

Predicted state:

$$\tilde{x}_{k/k-1} = F_k \widehat{x}_{k-1/k-1} + B_k u_{k-1} \tag{4.3}$$

Predicted estimate covariance:

$$P_{k/k-1} = F_k P_{k-1/k-1} F_k^{\mathrm{T}} + Q_{k-1} \tag{4.4}$$

Here, $\widehat{x}_{k-1/k-1}$ is the estimated state from the previous step, $\widetilde{x}_{k/k-1}$ is the predicted state of the current step, $P_{k/k-1}$ is the predicted covariance matrix of the current step, Q_{k-1} is the covariance of process noise w in discrete form, F_k and B_k are system dynamics and control distribution matrices, respectively, in discretized form, and u_{k-1} is the control vector . Subscript $k/k-1$ denotes that the computation is done in the current step k by using measurements of step $k-1$. Similarly, the subscript $k-1/k-1$ means that the variable is computed at step $k-1$ by taking measurements of step $k-1$ into consideration.

At the same time, Q_{k-1} should be introduced at that point. First, the continuous process noise covariance Q is:

$$Q = E\left(ww^{\mathrm{T}}\right) \tag{4.5}$$

where w is the process noise derived from a zero-mean Gaussian distribution and $E(.)$ is the operator of expected value.[1]

[1]In mathematics, the expected value of a random variable is the sum of the probabilities of each possible outcome multiplied by the numerical value of the outcome.

Then, the discrete process noise covariance matrix is:

$$Q_{k-1} = \int_{0}^{\Delta t} F(\tau)\, Q F^{\mathrm{T}}(\tau)\, \mathrm{d}\tau \tag{4.6}$$

Measurement Update The phase that the estimation of the present step is realized by using prediction phase outputs and, also, measurements information operation, where the predictions are improved.

Innovation or measurement residual:

$$\tilde{e}_k = z_k - H_k \tilde{x}_{k/k-1} \tag{4.7}$$

Innovation (or residual) covariance:

$$P_{\Delta k} = \left(H_k P_{k/k-1} H_k^{\mathrm{T}} + R_k\right) \tag{4.8}$$

Optimal Kalman gain:

$$K_k = P_{k/k-1} H_k^{\mathrm{T}} \left(H_k P_{k/k-1} H_k^{\mathrm{T}} + R_k\right)^{-1} \tag{4.9}$$

Updated state estimate:

$$\widehat{x}_{k/k} = \tilde{x}_{k/k-1} + K_k \tilde{e}_k \tag{4.10}$$

Updated estimate covariance:

$$P_{k/k} = (I - K_k H_k)\, P_{k/k-1} \tag{4.11}$$

Here, $\widetilde{e}_k$ is the innovation (or measurement residual), z_k is the measurement vector, H_k is the measurement matrix, K_k is the optimal Kalman gain, $\hat{x}_{k/k}$ is the estimated state, $P_{k/k}$ is the covariance matrix of the present step, and R_k is the covariance of measurement noise v in discrete form.

If discrete measurement noise covariance R_k is introduced, then similarly to Q_{k-1}:

$$R_k = E\left(v_k v_k^{\mathrm{T}}\right) \tag{4.12}$$

where v_k is the measurement white noise, which is assumed to be zero-mean Gaussian white noise.

Consequently, the extended Kalman filter equations are summarized in Table 4.1.

Table 4.1 Optimal Kalman filter equations

State prediction	$\tilde{x}_{k/k-1} = F_k\hat{x}_{k-1/k-1} + B_k u_{k-1}$
Covariance prediction	$P_{k/k-1} = F_k P_{k-1/k-1} F_k^{\mathrm{T}} + Q_{k-1}$
Innovation	$\tilde{e}_k = z_k - H_k\tilde{x}_{k/k-1}$
Innovation covariance	$P_{\Delta k} = \left(H_k P_{k/k-1} H_k^{\mathrm{T}} + R_k\right)$
Optimal Kalman gain	$K_k = P_{k/k-1} H_k^{\mathrm{T}}\left(H_k P_{k/k-1} H_k^{\mathrm{T}} + R_k\right)^{-1}$
State estimation	$\hat{x}_{k/k} = \tilde{x}_{k/k-1} + K_k\tilde{e}_k$
Covariance estimation	$P_{k/k} = (I - K_k H_k)\, P_{k/k-1}$

4.2.2 *Derivation of Optimal Kalman Gain*

The OKF uses a filter gain where the expected value of the square of the magnitude of error in posterior state estimation is minimized. In other words, the filter runs under some certain optimization law defined by the minimization rule of the indicated vector and, so, it has the optimal gain. If this optimal gain is modified in order to adapt the filter to the changing conditions, this means that the filter is no longer optimal and can be called an adaptive filter.

To make the meaning of optimal gain for an OKF more understandable, the gain derivation process can be described as the following [4].

First, it is known that the matrix $P_{k/k}$ is the covariance of estimation and, so:

$$P_{k/k} = \operatorname{cov}\left(x_k - \hat{x}_{k/k}\right) \tag{4.13}$$

where cov(·) is the covariance operator, x_k is the real value of the states in step k, and $\hat{x}_{k/k}$ is the estimation of the states in step k.

Then, by using the definition of state estimation $\hat{x}_{k/k}$ (Eq. 4.10):

$$P_{k/k} = \operatorname{cov}\left(x_k - \left(\tilde{x}_{k/k-1} + K_k\tilde{e}_k\right)\right) \tag{4.14}$$

substituting $\tilde{e}_k$ (Eq. 4.7):

$$P_{k/k} = \operatorname{cov}\left(x_k - \left(\tilde{x}_{k/k-1} + K_k\left(z_k - H_k\tilde{x}_{k/k-1}\right)\right)\right) \tag{4.15}$$

and z_k as:

$$z_k = H_k x_k + v_k \tag{4.16}$$

$$P_{k/k} = \operatorname{cov}\left(x_k - \left(\tilde{x}_{k/k-1} + K_k\left(H_k x_k + v_k - H_k\tilde{x}_{k/k-1}\right)\right)\right) \tag{4.17}$$

By collecting the necessary terms, the equation can also be written as:

$$P_{k/k} = \mathrm{cov}\left((I - K_k H_k)\left(x_k - \tilde{x}_{k/k-1}\right) - K_k v_k\right) \tag{4.18}$$

Since the measurement noise v_k is uncorrelated with other terms:

$$P_{k/k} = \mathrm{cov}\left((I - K_k H_k)\left(x_k - \tilde{x}_{k/k-1}\right)\right) + \mathrm{cov}\left(K_k v_k\right) \tag{4.19}$$

And by the properties of vector covariance:

$$P_{k/k} = (I - K_k H_k)\,\mathrm{cov}\left(x_k - \tilde{x}_{k/k-1}\right)(I - K_k H_k)^{\mathrm{T}} + K_k \mathrm{cov}\left(v_k\right) K_k^{\mathrm{T}} \tag{4.20}$$

As long as $\mathrm{cov}\left(x_k - \widetilde{x}_{k/k-1}\right)$ stands for $P_{k/k-1}$ and $\mathrm{cov}\left(v_k\right) = E\left(v_k v_k^{\mathrm{T}}\right) = R_k$:

$$\Rightarrow \quad P_{k/k} = (I - K_k H_k)\, P_{k/k-1}\, (I - K_k H_k)^{\mathrm{T}} + K_k R_k K_k^{\mathrm{T}} \tag{4.21}$$

This expression of updated covariance estimation is valid for every K_k and can be used for obtaining the optimal Kalman gain.

The Kalman filter is a minimum mean square error estimator and, as previously mentioned, to run the filter optimally, the minimum expected value of the square of the magnitude of state estimation is sought [4]. The estimation error is $x_k - \widehat{x}_{k/k}$ and the minimized quantity is $E\left[\left|x_k - \widehat{x}_{k/k}\right|^2\right]$. However, since $P_{k/k} = \mathrm{cov}\left(x_k - \widehat{x}_{k/k}\right)$, that is the same as minimizing the trace of covariance matrix $P_{k/k}$. It has been already found as:

$$\Rightarrow \quad P_{k/k} = (I - K_k H_k)\, P_{k/k-1}\, (I - K_k H_k)^{\mathrm{T}} + K_k R_k K_k^{\mathrm{T}}$$

If the terms are expended:

$$P_{k/k} = P_{k/k-1} - K_k H_k P_{k/k-1} - P_{k/k-1} K_k^{\mathrm{T}} H_k^{\mathrm{T}} + K_k P\Delta_k K_k^{\mathrm{T}} \tag{4.22}$$

where $P\Delta_k = \left(H_k P_{k/k-1} H_k^{\mathrm{T}} + R_k\right)$.

Therefore, for minimization:

$$\frac{\partial tr\left(P_{k/k}\right)}{\partial K_k} = -2\left(H_k P_{k/k-1}\right)^{\mathrm{T}} + 2K_k P\Delta_k = 0 \tag{4.23}$$

If Eq. (4.23) is solved for K_k, then:

$$K_k = P_{k/k-1} H_k^{\mathrm{T}} P\Delta_k^{\ -1} = P_{k/k-1} H_k^{\mathrm{T}} \left(H_k P_{k/k-1} H_k^{\mathrm{T}} + R_k\right)^{-1} \tag{4.24}$$

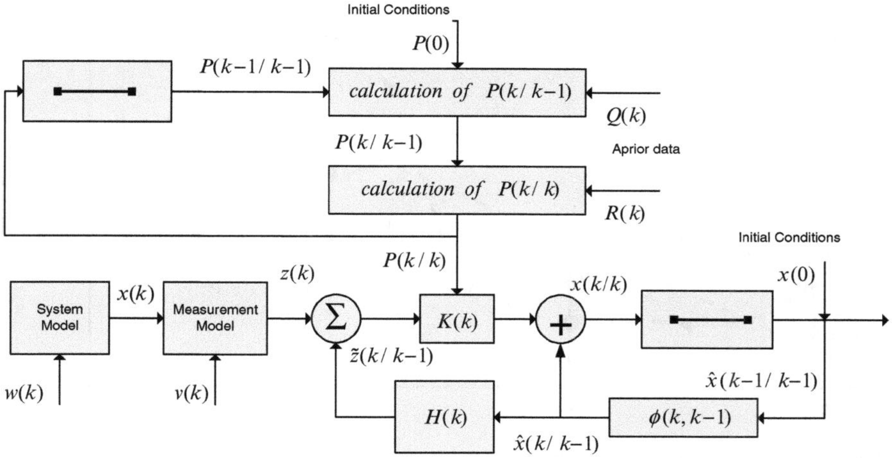

Fig. 4.1 Structural diagram of the Kalman filter

As can be seen, this equation is the same as Eq. (4.9) and the gain derived so far is called the optimal Kalman gain. Besides, if this expression of K_k is substituted into Eq. (4.22), then the equation:

$$P_{k/k} = (I - K_k H_k)\, P_{k/k-1}$$

can be found easily.

4.2.3 The Structure of the Kalman Filter

The structural diagram of the Kalman filter [5] is shown in Fig. 4.1.

The initial values $\bar{x}_0$ and P_0, correlation matrix of the system noise Q, and correlation matrix of the measurement noise R must be known previously in order for the filter to work.

According to formula (4.10), the estimation value is equal to the sum of the extrapolation value $\tilde{x}_{k/k-1}$ and the correction term $K_k \tilde{e}_k$. The extrapolation (prediction) value is calculated by multiplying the estimation value found in the previous step by the transfer matrix of the system. After that, correction is made to the extrapolation value. Hence, the Kalman filter works on the principle of correction of the prediction.

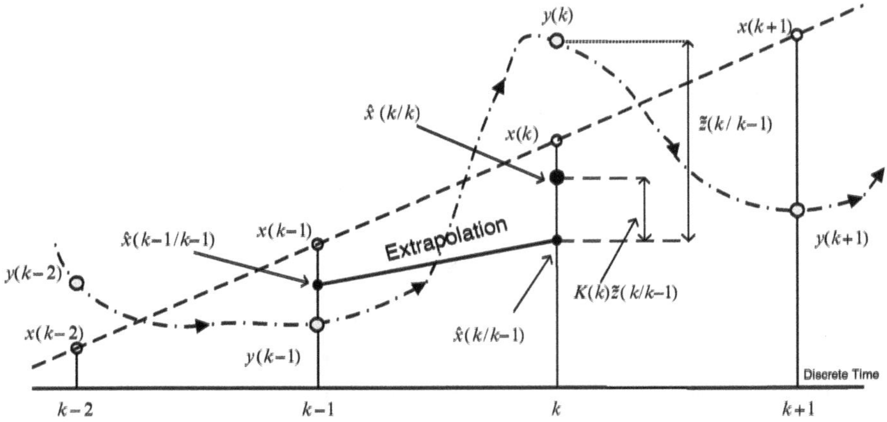

Fig. 4.2 Time-dependent diagram of the value estimation mechanism in the Kalman filter

The time-dependent diagram showing the mechanism of value generation in the Kalman filter is shown in Fig. 4.2. The filter algorithm includes the following operations:

1. One step further prediction of the value (determination of the extrapolation value) $\rightarrow \tilde{x}_{k/k-1}$
2. Multiplication of $\tilde{x}_{k/k-1}$ with H_k from the left; in other words, prediction of measurement
3. Determination of the difference between the measurement and the extrapolation value (innovation sequence) $\rightarrow \tilde{e}_k = z_k - H_k\tilde{x}_{k/k-1}$
4. Multiplication of $\tilde{e}_k$ with matrix K_k from the left and addition with $\tilde{x}_{k/k-1} \rightarrow \widehat{x}_{k/k}$
5. Storage of the $\widehat{x}_{k/k}$ value and repetition of the loop

It is postulated that the Kalman filter is the best filter among the subset of all linear filters and the best filter among the set of all filters when the noise processes are Gaussian [6].

The important features of the Kalman filter are given below [2]:

1. The filter can be easily realized using a computer once the model of the dynamic system is known
2. The value found by means of the filter is linear with respect to the measurement
3. The correlation matrix of the filtering error $P_{k/k}$ is not related to the measuring of z_k since the filter is linear, and it can be calculated beforehand
4. Filtering algorithms can be easily applied to the multidimensional case
5. The Kalman filter is equivalent to the Wiener filter in steady state for stationary dynamic filters

Adaptive filters become more important when the mathematical model of the system is not known previously or it is changed during the operation process.

This time evaluation operation is combined with the identification operation of the parameters and/or the model structure.

Nonlinear equations must be linearized first.

4.3 Stability of the Optimal Discrete Kalman Filter

As seen above, the Kalman filter is a time-domain filter, because the design is done in the time domain rather than the frequency domain. One of the advantages of the Kalman filter is its ability to estimate time-variable parameters. When the Kalman filter is used, the convergence of estimated values to the actual values of parameters depends on the stability condition of the filter. The stability of conventional digital filters is easily analyzed with z-transform methods [7]. We shall do the same for the optimum Kalman filter.

It is assumed that the optimum discrete Kalman filter [Eqs. (4.3), (4.4), (4.7), (4.8), (4.9), (4.10), and (4.11)] has reached a constant-gain condition. The estimation equation (4.10) can also be expressed for convenience:

$$\widehat{x}_{k/k} = \tilde{x}_{k/k-1} + K_k \left\{ z_k - H_k \tilde{x}_{k/k-1} \right\} \tag{4.25}$$

Replace $\tilde{x}_{k/k-1}$ with $F_k \widehat{x}_{k-1/k-1}$ in Eq. (4.25). After simplifications, the following equation is obtained:

$$\widehat{x}_{k/k} = \{F_k - K_k H_k F_k\} \widehat{x}_{k-1/k-1} + K_k z_k \tag{4.26}$$

Taking the z-transform of both sides of Eq. (4.26) and retarding $\widehat{x}_{k/k}$ by one step in the time domain is equivalent to multiplying $\widehat{X}^z_{k/k}$ by z^{-1} in the z-domain. In the z-domain, this means:

$$\widehat{X}^z_{k/k} = \{F_k - K_k H_k F_k\} z^{-1} \widehat{X}^z_{k/k} + K_k Z^z_k \tag{4.27}$$

After rearranging the terms, we have:

$$[zI - \{F_k - K_k H_k F_k\}] \widehat{X}^z_{k/k} = z K_k Z^z_k \tag{4.28}$$

where z denotes the usual z-transform variable and Z^z_k refers to the z-transform of the measurement vector.

It is well known from linear system theory that the bracketed quantity on the left side of Eq. (4.28) describes the natural modes of the system. The determinant of the bracketed $n \times n$ matrix gives the characteristic polynomial for the system; that is:

$$\text{Characteristic polynomial} = [zI - \{F_k - K_k H_k F_k\}] \,. \tag{4.29}$$

The roots of this polynomial provide information about the filter's stability. If all the roots lie inside the unit circle in the z-plane, the filter is stable; conversely, if any root lies on or outside the unit circle, the filter is unstable. As a matter of terminology, the roots of the characteristic polynomial are the same as the eigenvalues of $\{F_k - K_k H_k F_k\}$.

4.4 OKF for UAV State Estimation

When the Kalman filter algorithm is built for UAV state estimation, it is used for the combined longitudinal and lateral dynamics of aircraft. Hence, the state vector to be estimated is formed of $n = 9$ states as:

$$x = \begin{bmatrix} \Delta u & \Delta w & \Delta q & \Delta\theta & \Delta h & \Delta\beta & \Delta p & \Delta r & \Delta\phi \end{bmatrix}^{\mathrm{T}}, \tag{4.30}$$

while the control input vector is:

$$u = \begin{bmatrix} \Delta\delta_e & \Delta\delta_{\mathrm{T}} & \Delta\delta_a & \Delta\delta_r \end{bmatrix}^{\mathrm{T}}. \tag{4.31}$$

Here, Δu and Δw are the perturbations of the velocity components in the x- and z-directions, respectively, Δh is the perturbation of the height of the plane, $\Delta p, \Delta q, \Delta r$ are the perturbations of the angular velocities about the x, y, and z axes, respectively, $\Delta\theta, \Delta\phi, \Delta\beta$ are the perturbations of the pitch, roll, and sideslip angles, respectively, $\Delta\delta_e$, $\Delta\delta_a$, and $\Delta\delta_r$ are the perturbations of the elevator, aileron, and rudder deflections, respectively, and $\Delta\delta_{\mathrm{T}}$ is the change in the thrust.

Then, let us introduce the UAV process and observation models for the combined dynamics in the state space form as follows:

$$x_k = F_k x_{k-1} + B_k u_{k-1} + G_k w_k, \tag{4.32}$$

$$z_k = H_k x_k + v_k, \tag{4.33}$$

where F_k is the system dynamics matrix, B_k is the control distribution matrix, z_k is the measurement vector, G_k is the transition matrix of system noises, H_k is the measurement matrix, which is a 9×9 identity matrix in this case, and w_k and v_k are the white Gaussian system process and measurement noises, respectively:

$$E\left[w_k w_j^{\mathrm{T}}\right] = Q_k \delta_{kj}, \quad (4.33)$$

$$E\left[v_k v_j^{\mathrm{T}}\right] = R_k \delta_{kj}, \tag{4.34}$$

$$E\left[w_k v_j^{\mathrm{T}}\right] = 0. \tag{4.35}$$

Here, Q_k is the process noise covariance matrix, R_k is the measurement noise covariance matrix, and δ_{kj} is the Kronecker delta function.

After that, the optimal Kalman filter for such a combined UAV model can be obtained by following the steps given in Table 4.1.

4.5 Simulations

In this section, simulation results for the state estimation of the Zagi UAV are given. Simulation is realized in 500 steps for a period of 50 s with 0.1 s of sampling time, Δt. For a joined state vector of longitudinal and lateral equations (4.30), the measurement noise v is taken as:

$$v = N_r \times \left[0.92\ 0.92\ 0.083\ 0.17\ 0.2\ 0.05\ 0.083\ 0.083\ 0.17 \right]^{\mathrm{T}} \tag{4.36}$$

where N_r is a random number with zero mean and unit variance Gaussian distribution. Afterwards, R, the covariance of measurement noise, a 9×9 matrix for all steps, is given as:

$$R = \begin{bmatrix} 0.92^2 & \cdots & 0 \\ \vdots & \ddots & \vdots \\ 0 & \cdots & 0.17^2 \end{bmatrix} \tag{4.37}$$

For simulating measurements and also introducing system dynamics into the filter, the equations of motion of the UAV are put into discretized form. If the state space form of equations is shown as:

$$\dot{X} = AX + BU \tag{4.38}$$

the discretized equations become:

$$X_{i+1} = (I + \Delta t A)\, X_i + \Delta t B U_i \tag{4.39}$$

Here, A is the system matrix, B is the control distribution matrix, X is the state vector as given above, and U is the control vector, given as $u = \left[\Delta\delta_e\ \Delta\delta_{\mathrm{T}}\ \Delta\delta_a\ \Delta\delta_r \right]^{\mathrm{T}}$.

During simulations, the initial values of all states are assumed to be 0. On the other hand, the control vector is chosen as $u = \left[0.2\ 0\ 0\ 0.2 \right]^{\mathrm{T}}$rad.

Measurements are simulated by adding noise values to the real values of dynamic variations as:

$$Z_{i+1} = X_{i+1} + v \tag{4.40}$$

Besides for showing the drawbacks of the OKF and the necessity for filter adaptation in certain cases, errors are implemented in the measurements every 100th step. These abrupt errors, which represent a failure condition at the measurement channel, are formed by adding a constant term to the measurements of two states, $\Delta\theta$ and $\Delta\beta$.

Here, the simulation results are given for two states, Δu and $\Delta\theta$ in Figs. 4.3, 4.4, 4.5, 4.6, and 4.7. The results given in Figs. 4.5, 4.6, and 4.7 are for OKF in case of faulty measurements. Also Table 4.2 presents the estimation errors for all states at several time steps in case of faulty measurements. For any state, three graphs are introduced in a combined fashion: a plot that shows the Kalman estimation and simulated measurement values, a plot that gives the error of measurement and Kalman estimation according to the real value found by discretized dynamic equations (Eq. 4.39), and a plot that shows the variation of variance of Kalman estimations.

The simulation results show that the OKF gives sufficiently good estimation results in the case of normal operation of the estimation system (without faults). However, the optimal filter fails in providing accurate estimations in case of measurement faults, as expected. That signifies the importance of Kalman filter adaptation, which will be discussed in the next section.

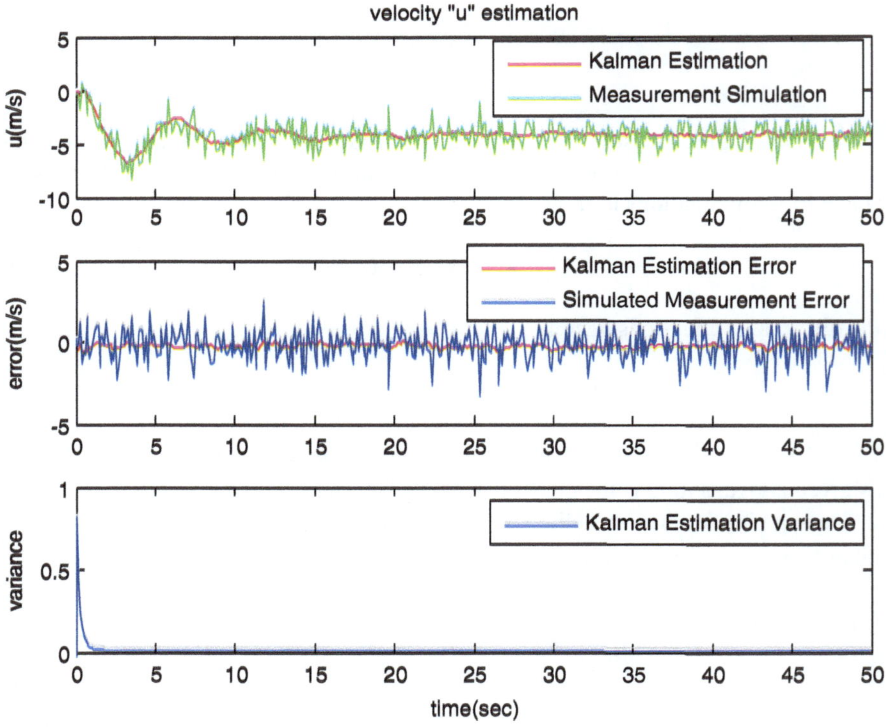

Fig. 4.3 *u* estimation of the optimal Kalman filter (OKF)

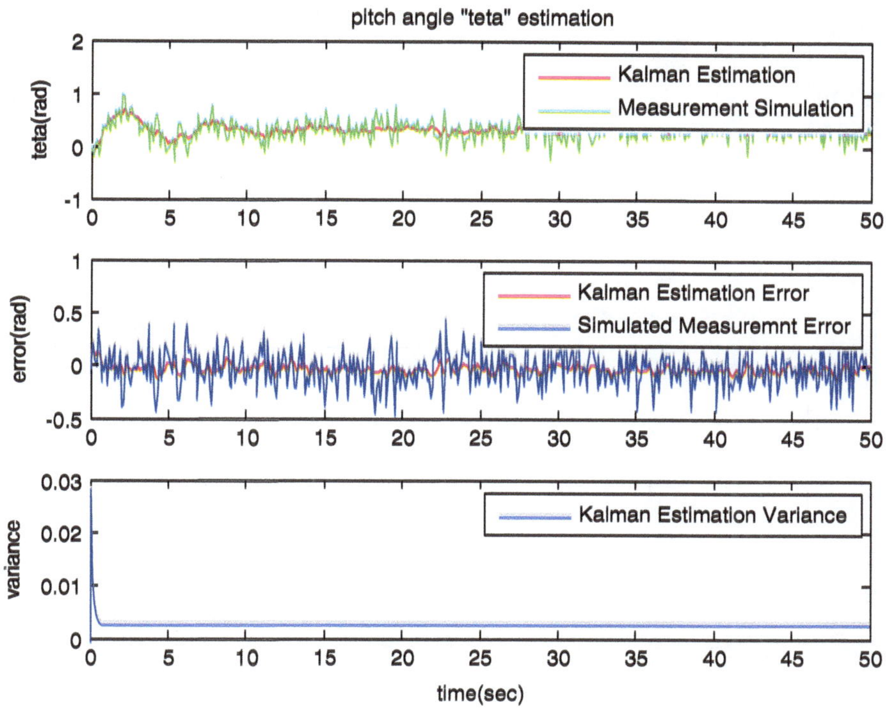

Fig. 4.4 θ estimation of the OKF

4.6 Necessity for Kalman Filter Adaptation

4.6.1 A Priori Uncertainty and Adaptation

The Kalman filter examined in the previous section is valid in the presence of whole prior statistical data. In practice, situations where a priori data are known with estimations or which exist completely are rarely encountered. For this case, the developed algorithms are not optimal and values estimated via these algorithms may not converge. Therefore, another approach is required to synthesize estimation algorithms.

A priori uncertainty degree can vary:

1. Exact a priori statistical uncertainty – In this case, the shape and parameters of the probability distribution rules of the components of measured and estimated random processes are not understood. However, allowable limited change areas of appropriate components of random processes are given. In such a case, the synthesis of estimation algorithms can be performed only on the basis of a guaranteed approach [8].

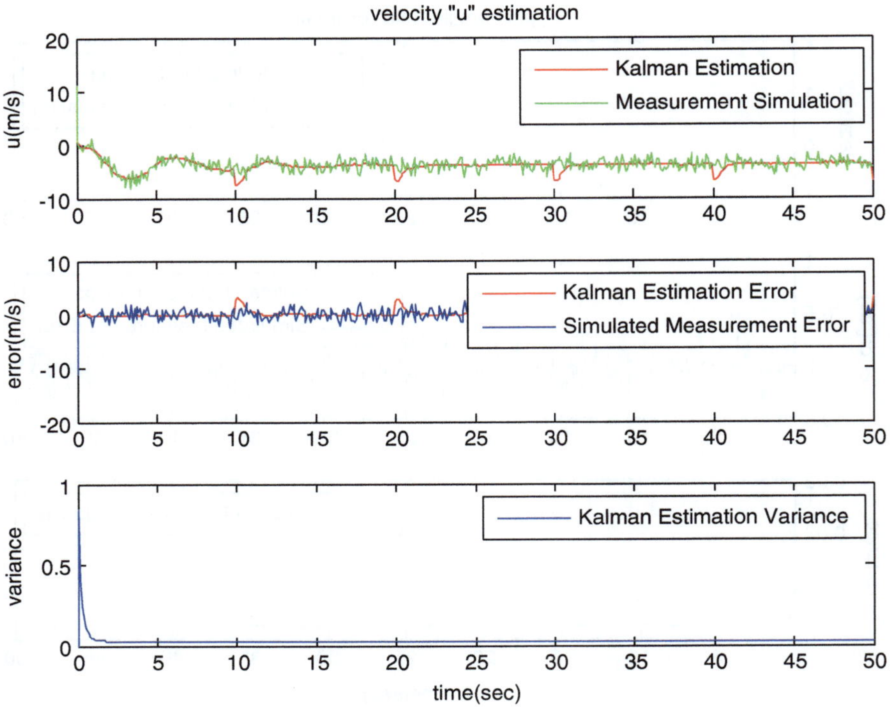

Fig. 4.5 *u* estimation of the OKF in case of faulty measurements

2. Partial prior statistical uncertainty – In this case, the probability distribution rules of some parts of measured and estimated random processes' components are known. The number of parameters which have unknown probability characteristics should not be too many. Increment in the number of these parameters causes the quality of the problem solution to decrease. On account of this, when there is a priori uncertainty in parameters, a distributions group is given instead of the probability distribution rule of random processes. The estimation algorithm from the distributions group must be chosen to provide optimal criterion. It means that the estimation algorithm should be adaptive.

There are three approaches for the adaptation problem: parametric approach, invariant principle-based approach, and structural approach. The parametric approach is the most commonly used. The parametric adaptive estimation algorithm is an algorithm that estimates the required components of random processes at the end of the measurement data transaction and restores a priori statistical characteristics of the dynamical system and measurements.

A self-tuning circuit is added to an ordinary Kalman filter in most parametric adaptive estimation algorithms. When the Kalman filter designed, it is supposed that the statistical characteristics of the system model, measurements, and noise

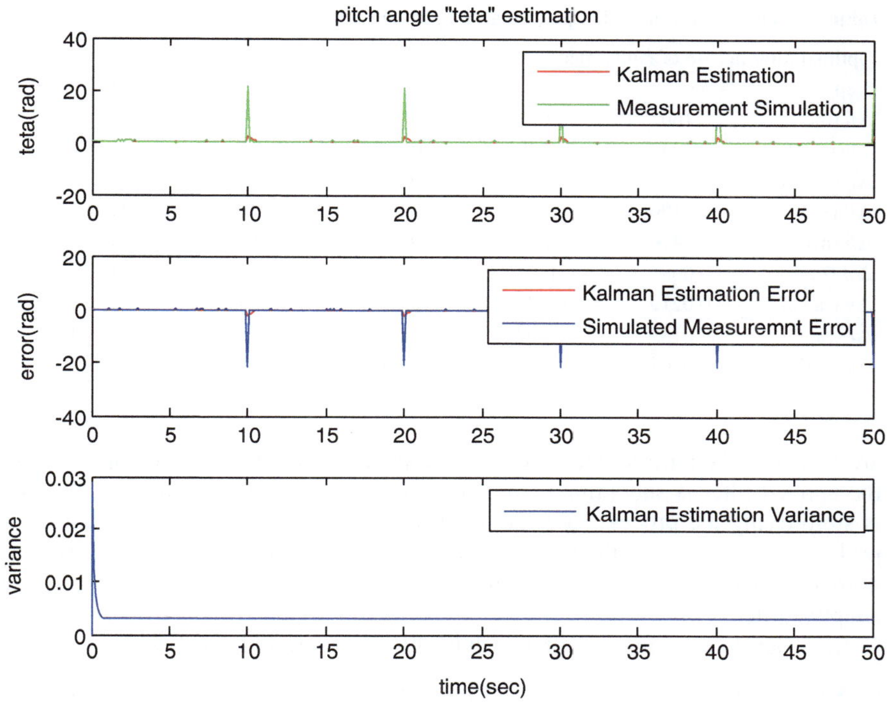

Fig. 4.6 θ estimation of the OKF in case of faulty measurements

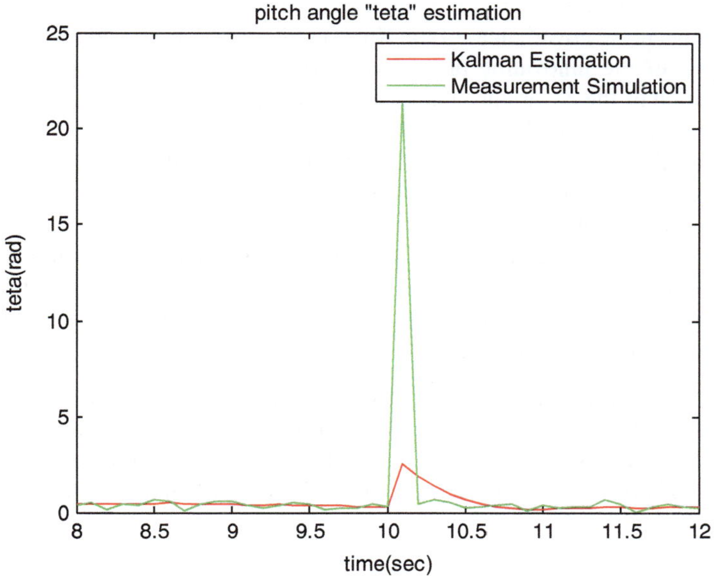

Fig. 4.7 Focused graph for θ estimation of the OKF in case of faulty measurements

Table 4.2 Optimal Kalman filter performance in case of faulty measurements

Optimal filter in case of faulty measurements					
Step	5 s.	20 s.	27.5 s.	40 s.	43 s.
Δu (m/s)	0.1602	3.0722	0.1089	2.9559	0.1362
Δw (m/s)	0.0259	0.8729	0.0326	0.9504	0.0499
Δq (rad/s)	0.0925	0.1222	0.0339	0.0325	0.0243
$\Delta \theta$ (rad)	0.0196	2.2815	0.0235	2.2582	0.0412
Δh (m)	0.1429	2.5808	0.0627	2.5163	0.04
$\Delta \beta$ (rad)	0.008	10.042	0.0023	10.016	0.0021
Δp (rad/s)	0.0219	0.4861	0.0258	0.5356	0.0227
Δr (rad/s)	0.0189	0.3699	0.0212	0.3121	0.0029
$\Delta \phi$ (rad)	0.0162	3.2214	0.0611	3.0618	0.0299

are known as being true. However, this supposition is not often accomplished, so the applied filter is suboptimal. Another reason for the suboptimality of the filter is simplifying the process calculations in the filter algorithm. In some cases, the problem of concretizing the fault caused by suboptimality appears. For this reason, noise covariances are estimated via the Kalman filter. This type of filter is called an adaptive filter.

Some approaches to designing adaptive filters are presented below:

- Multiple model-based adaptive estimation (MMAE). In the MMAE approach, adaptive filtering of a bank of Kalman filters runs in parallel under different models for the statistical filter information matrices, i.e., the process noise matrix Q and/or the update measurement noise matrix R.
- Unknown noise covariances of the Kalman filter are determined by the statistical analysis of the innovation or residual series. This might be either as a direct estimation of the covariance matrices or as an adaptation performed on the basis of covariance scaling [9] (Fig. 4.8).
- Noise estimation is performed via probability methods. Estimation values of unknown noise covariances are periodically used to renew the noise structure in filtering algorithms.
- By processing the same values via an iteration procedure, the determination of values of unknown covariances is based on the analysis of residual faults from the previous data processing circuit. This method is generally used when huge computational burden is required for a test.

Next, we review some of the existing common direct estimation techniques for Kalman filter adaptation. Newly proposed scaling-based techniques will be discussed in the next two chapters in detail. Readers are also referred to [9] for a comparison between these different methods for Kalman filter adaptation.

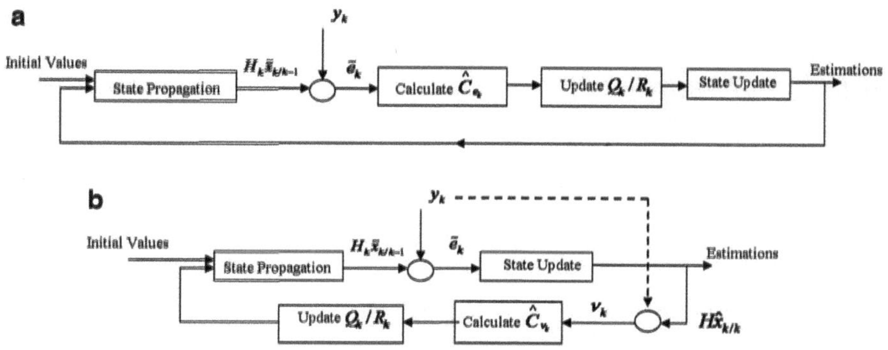

Fig. 4.8 The Kalman filter (KF) adaptation by the statistical analysis of the innovation and residual series: (**a**) innovation-based KF adaptation, (**b**) residual-based KF adaptation. For both adaptation techniques, updating of the process and measurement noise covariances might be either by estimation or by scaling

4.6.2 Innovation-Based Adaptive Estimation

Innovation-based adaptive estimation is the most widely used technique [10–15]. In this method, the appropriateness of the a priori characteristic of the noise to its real characteristic is determined by investigating the innovation sequence. In cases when the real innovation sequence differs from the white noise characteristics, estimation of the measurement noise (R) and system noise (Q) are realized. The analysis of innovation sequence statistics is simply checking for the sequence to verify that it has the expected zero value and the Gaussian white noise characteristic.

The covariance of the innovation sequence ($\tilde{e}_j$) is:

$$\hat{C}_{e_k} = \frac{1}{N}\sum_{j=k-N+1}^{k} \tilde{e}_j\tilde{e}_j^{\mathrm{T}} \tag{4.41}$$

where N is the dimension of the "sliding window".

If the real and a priori characteristics of the noise are different, R and Q must be determined consecutively and must also be included in the filtering algorithm. Thus, adaptive Kalman filters include filtering algorithms which are used to estimate system states, system parameters, and, also, the noise characteristics.

In innovation-based approaches, the previously predicted R and Q matrices are updated using the real-time measurements. The statistics matrices in the filter are updated using the formulas given below:

$$\hat{R}_k = \hat{C}_{e_k} - H_k P_{k/k-1} H_k^{\mathrm{T}} \tag{4.42}$$

$$\hat{Q}_k = K_k \hat{C}_{e_k} K_k^{\mathrm{T}} \tag{4.43}$$

4.6.3 Residual-Based Adaptive Estimation

In the residual-based approach [16, 17], the measurement covariance matrix and/or the system noise that are affected by the change in residual series are directly updated.

The residual series can be described by:

$$\nu_k = z_k - H\widehat{x}_{k/k}, \tag{4.44}$$

where z_k is the measurement vector, H is the measurement matrix, and $\widehat{x}_{k/k}$ is the estimation vector of the states.

The covariance matrix of the residual series is:

$$\widehat{C}_{\nu_k} = \frac{1}{N} \sum_{j=k-N+1}^{k} \nu_j \nu_j^{\mathrm{T}} \tag{4.45}$$

where N is the "sliding window" dimension.

The algorithms given below are used to estimate the R and Q matrices:

R matrix estimation algorithm:

$$\widehat{R}_k = \widehat{C}_{\nu_k} + H_k P_{k/k} H_k^{\mathrm{T}} \tag{4.46}$$

$P_{k/k}$ is the covariance matrix of the estimation error.

Q matrix estimation algorithm:

$$\widehat{Q}_k = \left(\frac{1}{N} \sum_{j=k-N+1}^{k} \Delta x_j \Delta x_j^{\mathrm{T}} \right) + P_{k/k} - FP_{k-1/k-1}F^{\mathrm{T}} \tag{4.47}$$

where $P_{k-1/k-1}$ is the covariance matrix of the estimation error in the previous step, F is the system transformation matrix, and Δx_k is the state correction series (the difference between the filter estimation and the extrapolation estimation):

$$\Delta x_k = \widehat{x}_{k/k} - \widehat{x}_{k/k-1} \tag{4.48}$$

In a stable state, only the first part of Eq. (4.47) is considered, in this case, taking into account that $\Delta x_k = K_k \nu_k$ in Eq. (4.47) can be approximated by Eq. (4.43).

It is understood that, for some practical applications [(like an integrated inertial navigation system (INS)/Global Positioning System (GPS)] [15], residual-based adaptive estimation methods are more effective than innovation-based approaches.

Some of the disadvantages of the described adaptive estimation methods are given below:

(a) In the estimation of unknown system and measurement noise covariance matrices using the residual- or innovation-based approaches, the innovation and residual vectors should be used for N cycles. This consumes a vast amount of computing power and makes determining appropriate sliding window dimension values a must.

(b) The measurement quantities, types, and distributions must be consistent in order to be able to use the residual- or innovation-based estimation methods. Otherwise, the covariance matrices of the system and measurement noise could not be estimated using the residual or innovation matrices.

(c) Innovation-based estimators may cause a negative covariance matrix if the value on the right-hand side of the formula $\widehat{R}_k = \hat{C}_{e_k} - H_k P_{k/k-1} H_k^{\mathrm{T}}$ is bigger than that on the left-hand side.

4.7 Conclusion

In this chapter, an OKF algorithm for the state estimation of a specific UAV's dynamics is presented and examined. Simulations realized by the use of a MATLAB program shows that the OKF algorithm gives accurate estimations under normal operation of the system (without faults). The OKF algorithm fails when faults occur in the estimation system.

Mathematically, the Kalman filter is a system of first-order ordinary differential equations with quadratic nonlinearities, which are solved on digital computers. When implementing navigational and other algorithms on a maintenance digital computer, incorrect results may occur due to equipment failure, malfunctioning of the computer, and noise during information transmission. For these cases, the adaptive Kalman filter can be used.

In accordance to the main motivation behind the study, supplying reliable parameter estimations to the control system of an autonomous UAV and ensuring that it completes its mission successfully is the objectives of this monograph.

References

1. Kalman RE (1960) A new approach to linear filtering and prediction problems. Trans ASME J Basic Eng 82:35–45
2. Hajiyev C, Caliskan F (2003) Fault diagnosis and reconfiguration in flight control systems. Kluwer Academic Publishers, Boston
3. Zarchan P, Musoff H (2000) Fundamentals of Kalman filtering: a practical approach. Progress in aeronautics and astronautics. AIAA, Reston
4. Sage AP, Melsa JL (1971) Estimation theory with applications in communication and control. McGraw-Hill, New York
5. Hajiyev CM (1999) Radio navigation. Istanbul Technical University, Istanbul (in Turkish)
6. Anderson BDO, Moore JB (1979) Optimal filtering. Prentice Hall, Englewood Cliffs
7. Brown RG, Hwang PYC (1997) Introduction to random signals and applied Kalman filtering with MATLAB exercises and solutions, 3rd edn. Wiley, New York

8. Ogarkov MA (1990) Methods for statistical estimation of random processes parameters. Energoatomizdat, Moscow (in Russian)
9. Hajiyev C, Soken HE (2015) Fault tolerant estimation of UAV dynamics via robust adaptive Kalman filter. In: Dimirovski GM (ed) Complex systems: relationships between control, communications and computing. Springer (in press), Berlin
10. Mehra RK (1970) On the identification of variances and adaptive Kalman filtering. IEEE Trans Autom Control 15(2):175–184
11. Mehra RK (1971) On-line identification of linear dynamic systems with applications to Kalman filtering. IEEE Trans Autom Control 16(1):12–21
12. Mehra RK (1971) Identification and adaptive Kalman filtering. Mechanic 3:34–52
13. Kailath T (1972) A note on least squares estimation by the innovations method. SIAM J Control 10(3):477–486
14. Maybeck PS (1982) Stochastic models, estimation, and control, vols I and II. Academic Press, New York
15. Salychev OS (1994) Special studies in dynamic estimation procedures with case studies in inertial surveying, ENGO 699.26 lecture notes. Department of Geomatics Engineering, University of Calgary, Canada
16. Mohamed AH, Schwarz KP (1999) Adaptive Kalman filtering for INS/GPS. J Geod 73:193–203
17. Wang J, Stewart MP, Tsakiri M (1999) Adaptive Kalman filtering for integration of GPS with GLONASS and INS. IUGG/IAG, Birmingham

Chapter 5
Estimation of Unmanned Aerial Vehicle Dynamics in the Presence of Sensor Faults

5.1 Introduction

When there are sufficient measurements, the necessary states of an unmanned aerial vehicle (UAV) can be estimated via a Kalman Filter. That is a desired procedure, since it is important to know precisely parameters like velocity, altitude, attitude, etc. When these states of the UAV are obtained without any problems, control of the aircraft can be achieved successfully. On the other hand, it is a process dependent upon the accuracy of measurements. If the measurements are not reliable due to any kind of malfunction in the estimation system, the filter gives inaccurate results and diverges over time. Since achieving fault tolerance in the design of a UAV flight control system is important, the filter should be built robustly in order to avoid such problems.

The Kalman filter approach to state estimation is quite sensitive to any measurement malfunctions (abnormal measurements, sudden shifts in the measurement channel, and other difficulties, such as decrease of instrument accuracy, an increase in background noise, etc.). If the conditions of operation of the measurement system do not correspond to the models used in the synthesis of the filter, then these changes resulting from some possible failures at the measurement channels significantly decrease the effectiveness of the estimation systems. When dealing with the measurement faults in previous estimation steps, rather than the current one, adaptive Kalman filters can be used to recover from the possible malfunctions.

A Kalman filter can be made adaptive and, hence, insensitive to the a priori measurements or system uncertainties by using various different techniques. The basic approaches to the adaptive Kalman filtering problem are multiple model-based adaptive estimation (MMAE) [1–3], innovation-based adaptive estimation (IAE) [4–6], and residual-based adaptive estimation (RAE) [7, 8]. While in the first approach a bank of Kalman filters run in parallel under different models for the filter's statistical information, in the other approaches, the adaptation is done

C. Hajiyev et al., *State Estimation and Control for Low-cost Unmanned Aerial Vehicles*, DOI 10.1007/978-3-319-16417-5_5

directly to the covariance matrices of the measurement and/or system noises based on the changes in the innovation or residual sequences.

In the methods described in [1–3], the faults are assumed to be known and the Kalman filters are designed for the known sensor/actuator faults. As the MMAE approach requires several parallel Kalman filters, and the faults should be known, it can only be used in a limited number of applications.

Estimation of the covariance matrices by IAE and RAE requires the usage of innovation vectors or residual vectors of m epoch. This increases the storage burden and presents the determination of the width of the moving window, m, as another problem. Furthermore, IAE and RAE estimators require that the number, type, and distribution of measurements for all epochs within a window should be consistent. If they are not, the covariance matrices of the measurement noises cannot be estimated based on the innovation or residual vectors.

The adaptive Kalman filter presented in [9] has been applied to fuse position signals from the Global Positioning System (GPS) and an inertial navigation system (INS) for autonomous mobile vehicles. The extended Kalman filter (EKF) and the noise characteristics have been modified using the fuzzy logic adaptive system. In [10], a method for multisensor data fusion based on the adaptive fuzzy Kalman filter is presented. This method is applied for fusing the position and orientation signals from the dead reckoning (DR) system and the GPS for landing vehicle navigation. The EKF and the characteristics of the measurement noise are modified using the fuzzy adaptive system, which is based on a covariance matching technique. It has been demonstrated that the fuzzy adaptive Kalman filter gives better results (more accurate) than the EKF [9, 10]. In [11], a fuzzy logic-based adaptive Kalman filter is used to build adaptive centralized, decentralized, and federated Kalman filters for adaptive multisensor data fusion. The adaptation carried out is in the sense of adaptively adjusting the measurement noise covariance matrix of each local filter to fit the actual statistics of the noise profiles present in the incoming measured data. A fuzzy inference system based on a covariance matching technique is used as the adaptation mechanism in the paper. The simulation results show that the architectures proposed by the authors are effective in situations where there are several sensors measuring the same parameters, but each one has different measurement dynamics and noise statistics. Although fuzzy logic-based adaptive Kalman filter algorithms perform well under specific circumstances, they are knowledge-based systems operating on linguistic variables, and these methods, which are based on human experiences, are not widely applicable to vital systems such as aircraft flight control systems.

Another concept is to scale the noise covariance matrix by multiplying it with a time-dependent variable. One of the methods for constructing such an algorithm is to use a single adaptive factor as a multiplier to the process or measurement noise covariance matrices [12–15]. This algorithm, which may be called an adaptive fading Kalman filter (AFKF), can be used when information about the dynamic process or the a priori measurements are absent [16]. However, when the point at issue is the current measurements, another technique to scale the measurement noise covariance matrix and make the filter robust (insensitive to measurement faults) should be proposed. Therefore, if there is a malfunction in the measurement system, a robust Kalman filter (RKF) algorithm can be utilized and, by the use of a measurement noise scale factor (MNSF) as a multiplier of the measurement noise covariance matrix, insensitiveness of the filter to the current measurement faults can be satisfied. Consequently, via a correction applied to the filter gain, good estimation behavior of the filter will be leveraged without being affected by faulty measurements [17].

However, the estimation performance of the Kalman filter differs for each variable when it is utilized for complex systems with multiple variables, and it may be not sufficient to use a single measurement noise scale factor (SMNSF) as a multiplier for the covariance matrices [18]. A single factor may not reflect corrective effects for the faulty measurement to the estimation process accurately. The technique that can be implemented to surmount this problem is to use multiple measurement noise scale factors (MMNSF) to fix the relevant components of the gain matrix individually. Unfortunately, thus far, any investigations about the comparison of the RKF with single and multiple scale factors have not been successful.

In this chapter, RKF algorithms with single and multiple MNSF, which require less computational burden for scale factor evaluation than the existing algorithms, are introduced and applied for the state estimation process of a UAV. The results of these algorithms are compared for different types of measurement malfunctions and recommendations about their utilization are given.

5.2 RKF with a Single Measurement Noise Scale Factor

Under normal operation conditions, where any kind of measurement malfunction is not observed, the optimal Kalman filter (OKF) gives sufficiently good estimation results. However, when the measurements are faulty due to malfunctions in the estimation system, such as abnormal measurements, sudden shifts, or step-like changes in the measurement channel, etc., filter estimation outputs become inaccurate.

Therefore, an RKF algorithm, which brings fault tolerance to the filter and secures accurate estimation results in case of faulty measurements without affecting the remaining good estimation characteristics, should be introduced.

The base of the RKF is the comparison of real and theoretical values of the covariance of the innovation sequence [17]. When the operational condition of the measurement system mismatches with the models used in the synthesis of the filter, then the Kalman filter gain changes according to the differentiation in the covariance matrix of the innovation sequence. Under these circumstances, the covariance matrix of the innovation sequence differs as [19, 20]:

$$P_{e_k} = H_k P_{k/k-1} H_k^{\mathrm{T}} + S_k R_k, \tag{5.1}$$

and, so, the Kalman gain becomes:

$$K_k = P_{k/k-1} H_k^{\mathrm{T}} \left(H_k P_{k/k-1} H_k^{\mathrm{T}} + S_k R_k \right)^{-1}. \tag{5.2}$$

Here, S_k is the SMNSF.

Due to this approach, the Kalman gain is changed when the predicted observation $H_k \tilde{x}_{k/k-1}$ is considerably different from the actual observation y_k because of the significant changes in the operational condition of the measurement system. In other words, if the real value of the filtration error exceeds the theoretical error:

$$tr\left\{ \tilde{e}_k \tilde{e}_k^{\mathrm{T}} \right\} \geq tr\left\{ H_k P_{k/k-1} H_k^{\mathrm{T}} + R_k \right\}, \tag{5.3}$$

the filter must be run robustly. Here, $tr\{\cdot\}$ denotes the trace of the related matrix.

In order to determine the scale factor, S_k, let us substitute Eq. (5.1) into Eq. (5.3) and use as input the scale factor calculated at the point where condition (5.3) is satisfied:

$$tr\left\{ \tilde{e}_k \tilde{e}_k^{\mathrm{T}} \right\} = tr\left\{ H_k P_{k/k-1} H_k^{\mathrm{T}} \right\} + S_k tr\left\{ R_k \right\}. \tag{5.4}$$

Then, in light of the equality $tr\left\{ \tilde{e}_k \tilde{e}_k^{\mathrm{T}} \right\} = \tilde{e}_k^{\mathrm{T}} \tilde{e}_k$, S_k can be written as:

$$S_k = \frac{\tilde{e}_k^{\mathrm{T}} \tilde{e}_k - tr\left\{ H_k P_{k/k-1} H_k^{\mathrm{T}} \right\}}{tr\left\{ R_k \right\}}. \tag{5.5}$$

In case of malfunction in the measurement system, the adaptation of the Kalman filter is performed via automatically correcting the Kalman gain. If the condition in Eq. (5.3) is met, then it brings about an increase in the scale factor S_k. A larger S_k causes a smaller Kalman gain (5.2) because of the covariance of the innovation sequence (5.1), which is also increased in the robust case. Consequently, a small Kalman gain value reduces the effect of the faulty innovation sequence on the state estimation process. In all other cases, where the measurement system operates normally, the SMNSF takes the value of $S_k = 1$ and, so, the filter runs optimally.

It must be noted that, due to the scale factor S_k, the covariance of the estimation error of the RKF increases in comparison with the OKF. Therefore, the robust algorithm is used only when the measurements are faulty and, in all other cases, the procedure is run optimally with the regular Kalman filter. The process is controlled by the use of a kind of statistical information. At that point, the two following hypotheses may be introduced:

γ_0; the system is operating normally
γ_1; there is a malfunction in the estimation system

To detect failures, a statistical function may be defined as [21]:

$$\beta_k = \tilde{e}_k^{\mathrm{T}} \left[H_k P_{k/k-1} H_k^{\mathrm{T}} + R_k \right]^{-1} \tilde{e}_k. \tag{5.6}$$

This statistical function has a χ^2 distribution with M degrees of freedom, where M is the dimension of the innovation vector.

If the level of significance, α, is selected as:

$$P\left\{\chi^2 > \chi^2_{\alpha,M}\right\} = \alpha; \quad 0 < \alpha < 1, \tag{5.7}$$

the threshold value, $\chi^2_{\alpha,M}$, can be found. Hence, when the hypothesis γ_1 is correct, the statistical value of β_k will be greater than the threshold value $\chi^2_{\alpha,M}$, i.e.:

$$\begin{array}{ll} \gamma_0 : \beta_k \leq \chi^2_{\alpha,M} & \forall k \\ \gamma_1 : \beta_k > \chi^2_{\alpha,M} & \exists k. \end{array} \tag{5.8}$$

5.3 RKF with Multiple Measurement Noise Scale Factors

As previously mentioned, it is possible to adapt the filter by using a single scale factor as a corrective term on the filter gain [17, 19], but this is not a safe procedure as long as the filter performance differs for each state for complex systems with multiple variables [18, 22]. The preferred method is to use a matrix built from multiple measurement noise scale factors to fix the relevant terms of the measurement noise covariance matrix, and, consequently, the Kalman gain.

In order to determine the scale matrix, an innovation-based process may be followed. It is known that the Kalman filter innovation sequence can be determined by Eq. (5.3). Then, as the next step, real and theoretical values of the innovation covariance matrix must be compared. When there is a measurement malfunction in the estimation system, the real error will exceed the theoretical error. Hence, if a scale matrix, S_k, is added into the algorithm as [23] follows:

$$\frac{1}{\mu}\sum_{j=k-\mu+1}^{k} \tilde{e}_j\tilde{e}_j^{\mathrm{T}} = H_k P_{k/k-1} H_k^{\mathrm{T}} + S_k R_k, \tag{5.9}$$

then it can be determined using the formula:

$$S_k = \left(\frac{1}{\mu}\sum_{j=k-\mu+1}^{k} \tilde{e}_j\tilde{e}_j^{\mathrm{T}} - H_k P_{k/k-1} H_k^{\mathrm{T}}\right) R_k^{-1}. \tag{5.10}$$

Here, μ is the width of the moving window.

In the case of normal operation, the scale matrix will be a unit matrix, $S_k = I$. Here, I represents the unit matrix. Nonetheless, as μ is a limited number due to the quantity of measurements and computations performed using a computer, it implies errors such as approximation errors and rounding off errors; the matrix S_k, found by the use of Eq. (5.10), may not be diagonal and may have diagonal elements which are "negative" or less than "one" (actually, that is physically impossible).

Therefore, in order to avoid such situations, it is suggested that the scale matrix is composed by the following rule:

$$S^* = \mathrm{diag}\left(s_1^*, s_2^*, \ldots, s_n^*\right), \tag{5.11}$$

where:

$$s_i^* = \max\left\{1, S_{ii}\right\} \quad i = 1, n. \tag{5.12}$$

Here, S_{ii} represents the ith diagonal element of the matrix S. Apart from that point, if the measurements are faulty, S_k^* will change and, so, affect the Kalman gain matrix:

$$K_k = P_{k/k-1} H_k^{\mathrm{T}} \left(H_k P_{k/k-1} H_k^{\mathrm{T}} + S_k^* R_k\right)^{-1}. \tag{5.13}$$

Therefore, in case of any kind of malfunctions, related elements of the scale matrix, which correspond to the faulty component of the measurement vector, increase, and that brings about a smaller Kalman gain, which reduces the effect of the innovation on the state update process. As a result, more accurate estimation results can be obtained.

As long as the proposed RKF is not optimal due to the scale matrix S_k^*, the robust algorithm is again operated only when the measurements are faulty and, in all other cases, the procedure is run optimally with the regular OKF. The same statistical information as the SMNSF-based algorithm, which is defined by Eqs. (5.6), (5.7), and (5.8), is used as the supervision criteria for starting the adaptation of the RKF.

5.4 Comparison of the *R*-adaptation Techniques

The *R*-adaptation procedures with single and multiple measurement noise scale factors are compared through simulations. Simulations are realized in 1,000 steps for a period of 100 s with 0.1 s of sampling time, Δt. As an experimental platform, the Zagi UAV is chosen and Kalman filter applications are performed with respect to the dynamics and characteristics of this UAV.

During simulations, for testing the RKF algorithm, four kinds of measurement malfunction scenarios are considered: instantaneous abnormal measurements, continuous bias, measurement noise increment, and fault of zero output. Also, in case of measurement faults, simulations are also performed for the OKF to compare the results with the RKF algorithms and understand the efficiency of the RKF in more detail. Nonetheless, $\chi^2_{\alpha,s}$ is taken as 16.919, and this value comes from a Chi-square distribution when the number of degrees of freedom is 9 and the reliability level is 95 %.

In this chapter, the first set of figures gives the OKF or RKF state estimation results and the actual values are compared. The second set of figures shows the error of the estimation process based on the actual state values of the UAV. The last set of figures indicates the variance of the estimation that corresponds to the related diagonal element of the covariance matrix of estimation errors.

5.4.1 Instantaneous Abnormal Measurements

Instantaneous abnormal measurements are simulated by adding a constant term to the pitch angle, $\Delta\theta$, with measurement every 20th second. It is obvious from Figs. 5.1, 5.2, and 5.3 that both RKF algorithms (with SMNSF and MMNSF) give more accurate estimation results than the OKF.

The results obtained by the regular OKF are not reliable when the obtained measurements contain an error. However, the RKF with SMNSF and MMNSF maintain their estimation characteristic for the whole process and affords precise estimation outputs in case of abnormal measurements, as well as normal operational conditions.

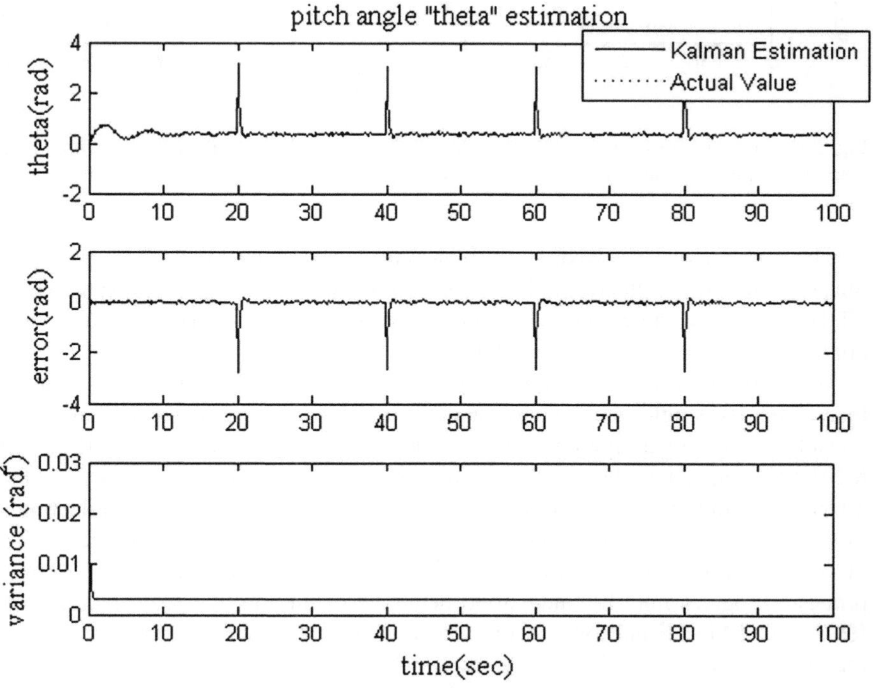

Fig. 5.1 Estimation result for the pitch angle θ by the regular optimal Kalman filter (OKF) in case of instantaneous abnormal measurements

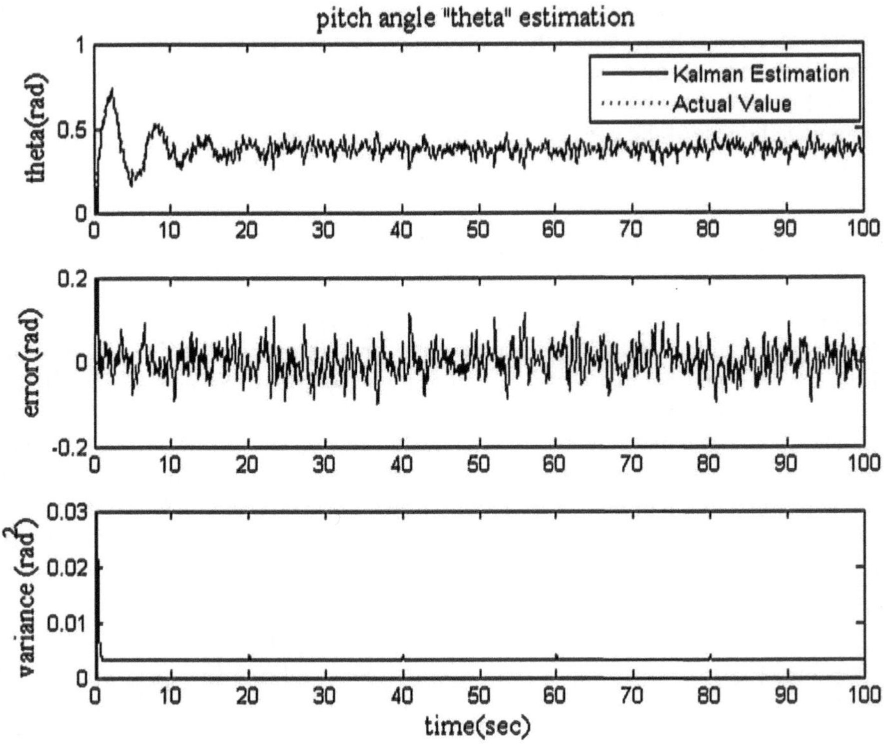

Fig. 5.2 Estimation result for the pitch angle θ by the robust Kalman filter (RKF) with a single measurement noise scale factor (SMNSF) in case of instantaneous abnormal measurements

The superiority of the RKF algorithms may also be seen in Table 5.1, which gives the absolute estimation errors of the filters for the indicated time step. Note that, for all the tables given hereafter, the highlighted results are obtained at time steps where abnormal measurements are implemented.

When the measurements are faulty, the RKF with SMNSF compensates by increasing its single scale factor and all of the measurements are taken into consideration with a small weight for these time steps. Besides, the RKF with MMNSF ensures the robustness of the filter by increasing related scale factors of the matrix individually. Increment of the related scale factors brings about a decrement in the related components of the Kalman gain, so as to reduce the corrective effect of the innovation sequences of the faulty measurements on the state estimation process. Progress may be understood better by examining the scale factor of the RKF with SMNSF and one of the scalar measures (trace, determinant, and the maximum eigenvalue) of the scale matrix of the RKF with MMNSF. The trace of the scale matrix is examined in Table 5.2 as a scalar measure. Similar results have been obtained when the fault is implemented in the other measurement channels.

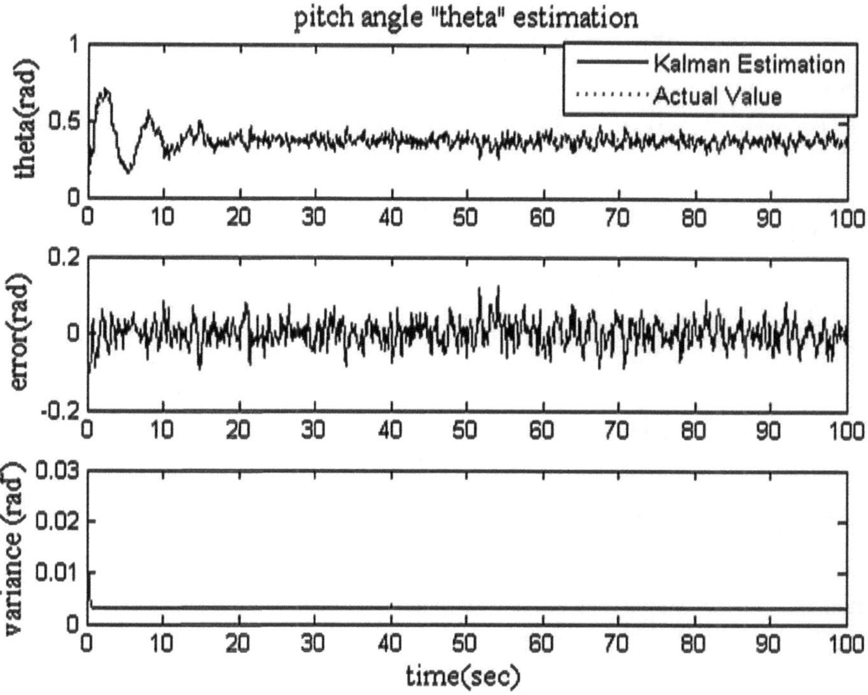

Fig. 5.3 Estimation result for the pitch angle θ by the RKF with multiple measurement noise scale factors (MMNSF) in case of instantaneous abnormal measurements

Even though both of the RKFs give similar estimation results in case of instantaneous abnormal measurements, the RKF with MMNSF can be thought of as a more advantageous algorithm, as it takes the faulty measurements into account individually. Disregarding all of the measurements, as the RKF with SMNSF does, affects the estimation procedure of all the states. However, since the abnormal measurement is implemented in only one measurement channel, leaving out the related measurement would be more significant, especially when the filter mathematical model of the UAV, which is constructed with the decoupled longitudinal and lateral dynamics, is taken into consideration.

5.4.2 Continuous Bias of Measurements

The continuous bias term is formed by adding a constant term to the measurement of pitch angle, $\Delta\theta$, in between the 30th and 60th seconds. As Fig. 5.4 shows, again, the OKF fails in estimating states accurately. In contrast, the RKF with SMNSF derogates estimation errors to a perceptible degree and gives relatively better estimation outputs (Fig. 5.5). In this case, the RKF with MMNSF gives

Table 5.1 Comparison of absolute estimation errors of regular the optimal Kalman filter (OKF), robust Kalman filter (RKF) with a single measurement noise scale factor (SMNSF), and RKF with multiple measurement noise scale factors (MMNSF) in case of instantaneous abnormal measurements

Parameter	Absolute values of error for regular OKF		Absolute values of error for RKF with SMNSF		Absolute values of error for RKF with MMNSF	
	40 s	70 s	40 s	70 s	40 s	70 s
Δu(m/s)	3.4509	0.0609	0.0738	0.1009	0.0746	0.0093
Δw(m/s)	1.0658	0.0715	0.0419	0.177	0.0148	0.009
Δq(rad/s)	0.1541	0.0402	0.0011	0.1155	0.0905	0.0407
$\Delta\theta$(rad)	2.6624	0.0139	0.003	0.0504	0.0165	0.0073
Δh(m)	2.9367	0.0266	0.0203	0.1519	0.0492	0.0075
$\Delta\beta$(rad)	0.0318	0.0283	0.0448	0.0015	0.0137	0.0145
Δp(rad/s)	0.006	0.0009	0.0047	0.0193	0.0894	0.0833
Δr(rad /s)	0.0085	0.0587	0.0424	0.0241	0.0195	0.0049
$\Delta\phi$(rad)	0.04	0.0931	0.1455	0.0156	0.0037	0.0606

Table 5.2 MNSF investigation in case of instantaneous abnormal measurements

Parameter	20 s	40 s	45 s	50 s	60 s	90 s
Scale factor of RKF with SMNSF S_k	346	345	1	1	343	1
Trace of scale matrix of RKF with MMNSF $\mathrm{trace}\left(S_k^*\right)$	1,094	1,097	1	1	1,095	1

superior estimation results to the RKF with SMNSF by reducing the effect of the innovation term of $\Delta\theta$ measurement and totally eliminating the estimation error caused by the biased measurements of this channel (Fig. 5.6). As it disregards only the related measurements, by the use of the innovative terms taken from all other sensors operating correctly, it ensures accurate estimation outputs throughout this period. It is possible to obtain the same results when the continuous bias is implemented in the other measurement channels.

Comparison of the performance of the regular OKF, RKF with SMNSF, and RKF with MMNSF in case of continuous bias of $\Delta\theta$ measurement can be seen in Table 5.3.

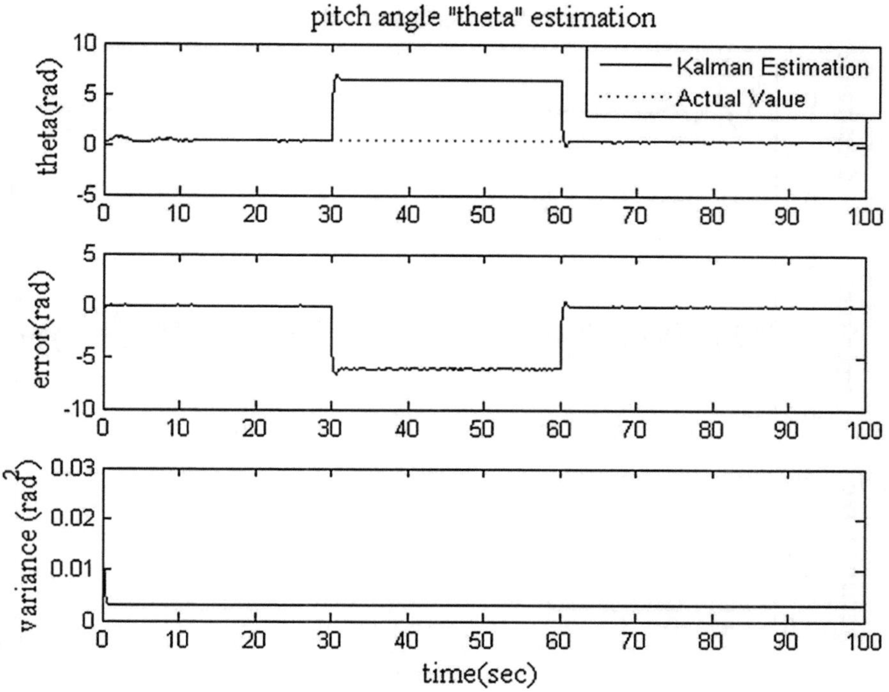

Fig. 5.4 Estimation result for the pitch angle θ by the regular OKF in case of continuous bias

In this table, the time means of absolute error values are given and they are calculated for 30 s (between the 30th and 60th seconds for the fault case and between the 65th and 95th seconds for the no fault case).

Again, it is possible to investigate the behavior of scale factors of both RKFs (Table 5.4).

5.4.3 Measurement Noise Increment

In the third measurement malfunction scenario, measurement fault is characterized by multiplying the variance of the noise of the velocity component Δu measurement with a constant term between the 30th and 60th seconds. Figures 5.7, 5.8, and 5.9 show that the OKF outputs involve error, while the RKF algorithms achieve estimation of the states accurately. In this case, it is not possible to detect the superiority of the RKF with MMNSF by examining the estimation results, although it is a known fact that individual increase of related scale factors, which corresponds to the faulty measurements, is advantageous compared to disregarding

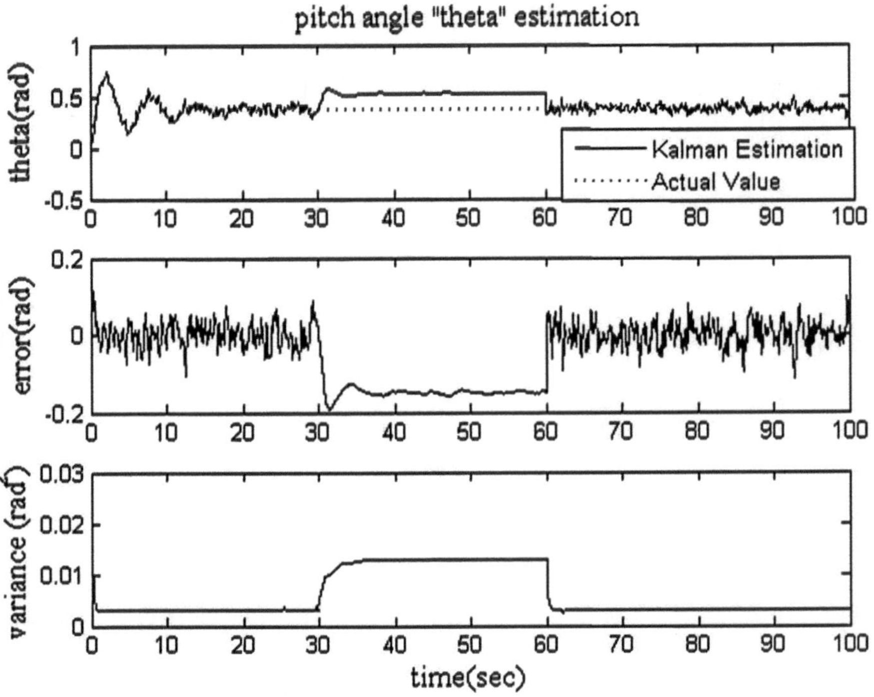

Fig. 5.5 Estimation result for the pitch angle θ by the RKF with SMNSF in case of continuous bias

all measurements at the same time. Simulations give similar results when the noise increment is implemented in the other measurement channels.

Table 5.5 shows a comparison of the mean of absolute estimation errors of the regular OKF, RKF with SMNSF, and RKF with MMNSF in case of measurement noise increment.

Table 5.6 presents the variation of the scale factor and the scalar measures of the scale matrix.

5.4.4 *Fault of Zero Output*

In the last measurement malfunction scenario, it is assumed that height, Δh, cannot be measured and the related sensor gives "0" as the output. This sort of fault is easily simulated by taking the Δh measurement as "0" for the filter algorithm in between the 30th and 60th seconds. As shown in Fig. 5.10, the OKF fails in estimating the height and gives a result of "0" in light of the measurement. Similarly, despite working robustly at the initial steps of the measurement fault implementation

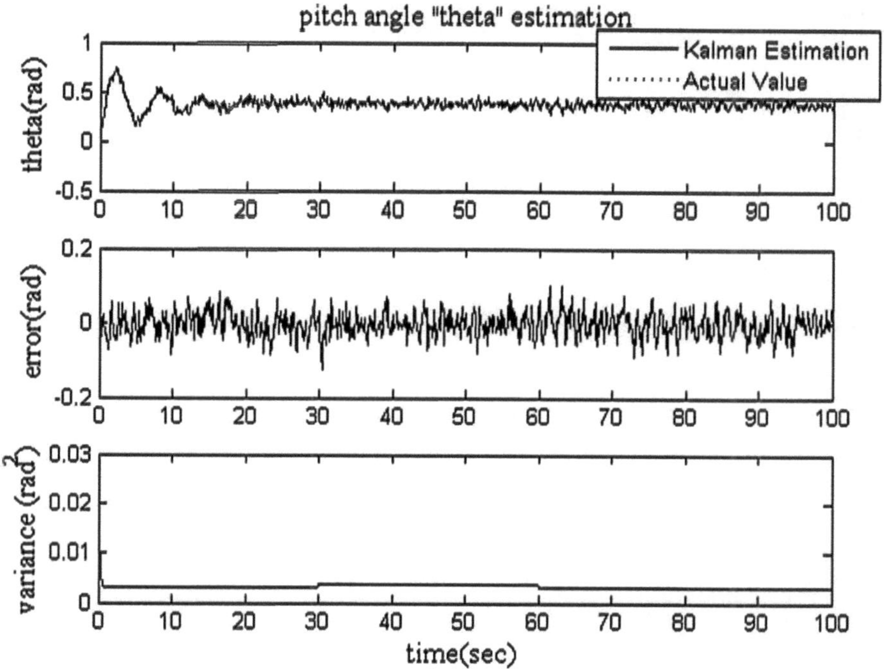

Fig. 5.6 Estimation result for the pitch angle θ by the RKF with MMNSF in case of continuous bias

Table 5.3 Comparison of the mean absolute estimation errors of the regular OKF, RKF with SMNSF, and RKF with MMNSF in case of continuous bias of measurements (for 30 s)

Parameter	Absolute values of error for regular OKF		Absolute values of error for RKF with SMNSF		Absolute values of error for RKF with MMNSF	
	Fault	No fault	Fault	No fault	Fault	No fault
Δu(m/s)	23.091	0.0906	1.9103	0.0982	0.0846	0.1066
Δw(m/s)	6.9494	0.0605	0.3337	0.0603	0.0637	0.06
Δq(rad/s)	0.7806	0.0535	0.2153	0.0518	0.0512	0.0515
$\Delta\theta$(rad)	6.0079	0.027	0.1473	0.0261	0.0225	0.0276
Δh(m)	12.9792	0.0792	2.6621	0.0855	0.082	0.086
$\Delta\beta$(rad)	0.0248	0.0227	0.0083	0.0212	0.023	0.0215
Δp(rad/s)	0.0314	0.0327	0.0016	0.0311	0.0267	0.0332
Δr(rad/s)	0.0238	0.0255	0.0078	0.0233	0.0219	0.0249
$\Delta\phi$(rad)	0.0424	0.0441	0.009	0.0375	0.0327	0.0403

Table 5.4 MNSF investigation in case of continuous bias of measurements

Parameter	20 s	40 s	45 s	50 s	60 s	90 s
Scale factor of RKF with SMNSF S_k	1	290	303	294	297	1
Trace of scale matrix of RKF with MMNSF trace$\left(S_k^*\right)$	1	6,583	9,857	13,123	18,294	1

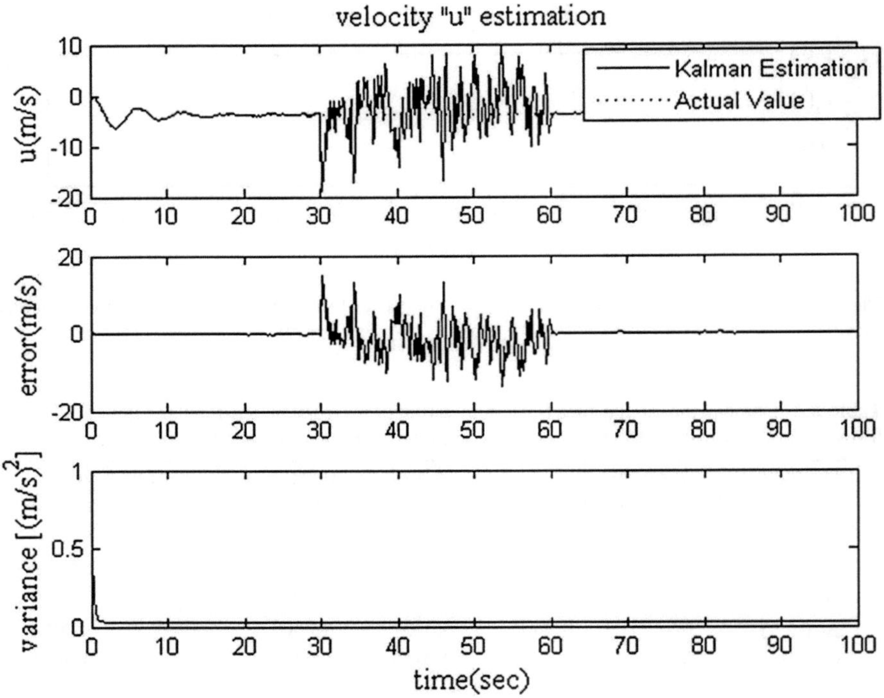

Fig. 5.7 Estimation result for the velocity component u by the regular OKF in case of measurement noise increment

(between the 30th and 40th seconds approximately), the RKF with SMNSF accepts the Δh measurement as a true value and estimates it as "0" (Fig. 5.11).

Clearly, not taking healthy measurements into consideration throughout this period affects the estimation procedure of the RKF with SMNSF. However, the RKF

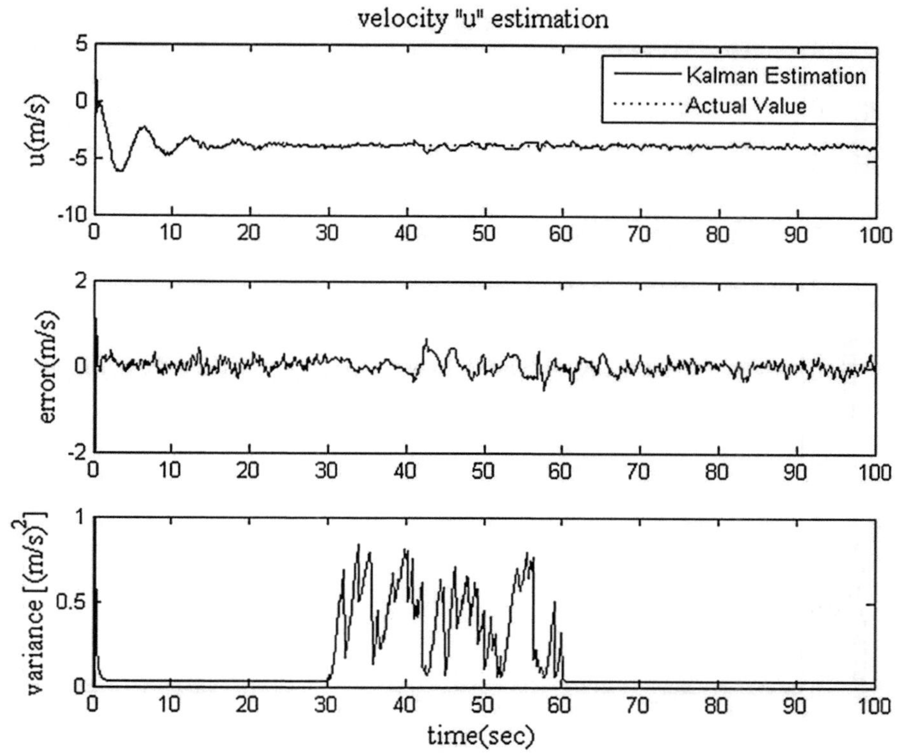

Fig. 5.8 Estimation result for the velocity component u by the RKF with SMNSF in case of measurement noise increment

with MMNSF performs well under these circumstances and, by disregarding only the faulty Δh measurement and using innovative terms taken from all other sensors operating correctly, accurately estimates all of the states (Fig. 5.12). This behavior can also be seen from Table 5.7.

Nonetheless, the behavior of the RKFs can be observed by investigating the variation of scale factor and the scalar measures of the scale matrix (Table 5.8). As can be seen, the scale factor takes a higher value than one only at the initial steps of the malfunction scenario, while the matrix scalar measures conserve that increment for the whole process.

5.5 Remark on Stability

Indeed, the proposed RKFs are not so different from the linear Kalman filter from the point of view of structure; they may be assumed to be a modification. In this sense, a similar approach to that of a linear Kalman filter may be followed

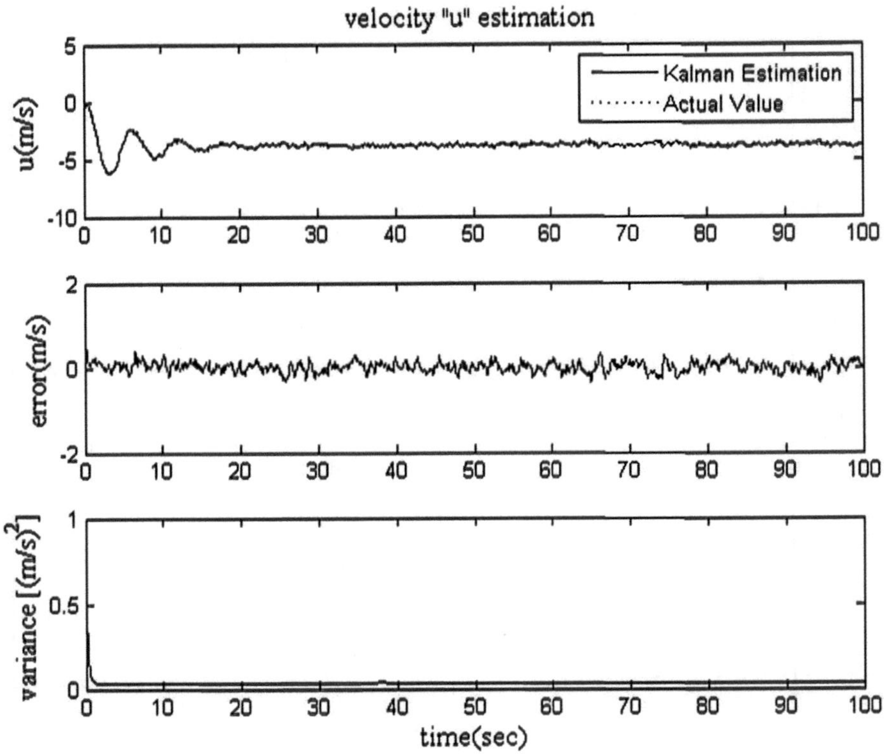

Fig. 5.9 Estimation result for the velocity component u by the RKF with MMNSF in case of measurement noise increment

Table 5.5 Comparison of the mean absolute estimation errors of the regular OKF, RKF with SMNSF, and RKF with MMNSF in case of measurement noise increment (for 30 s)

Parameter	Absolute values of error for regular OKF		Absolute values of error for RKFwith SMNSF		Absolute values of error for RKF with MMNSF	
	Fault	No fault	Fault	No fault	Fault	No fault
Δu(m/s)	4.0183	0.0901	0.1571	0.0959	0.0836	0.112
Δw(m/s)	1.2699	0.0656	0.0302	0.0644	0.052	0.0656
Δq(rad/s)	0.1462	0.056	0.0197	0.0562	0.0524	0.0557
$\Delta \theta$(rad)	0.4319	0.0287	0.0177	0.0266	0.0243	0.0275
Δh(m)	1.336	0.0869	0.1411	0.0913	0.0818	0.0905
$\Delta \beta$(rad)	0.0227	0.0226	0.0328	0.0222	0.0225	0.0224
Δp(rad/s)	0.0295	0.0314	0.0125	0.0308	0.0283	0.0316
Δr(rad/s)	0.0233	0.0231	0.03	0.0214	0.0198	0.0221
$\Delta \phi$(rad)	0.0455	0.0344	0.0443	0.0355	0.0388	0.0395

Table 5.6 MNSF investigation in case of measurement noise increment

Parameter	20 s	40 s	45 s	50 s	60 s	90 s
Scale factor of RKF with SMNSF S_k	1	279	1	254	6308	1
Trace of scale matrix of RKF with MMNSF trace$\left(S_k^*\right)$	1	9,457	3,973	18,241	4,935	1

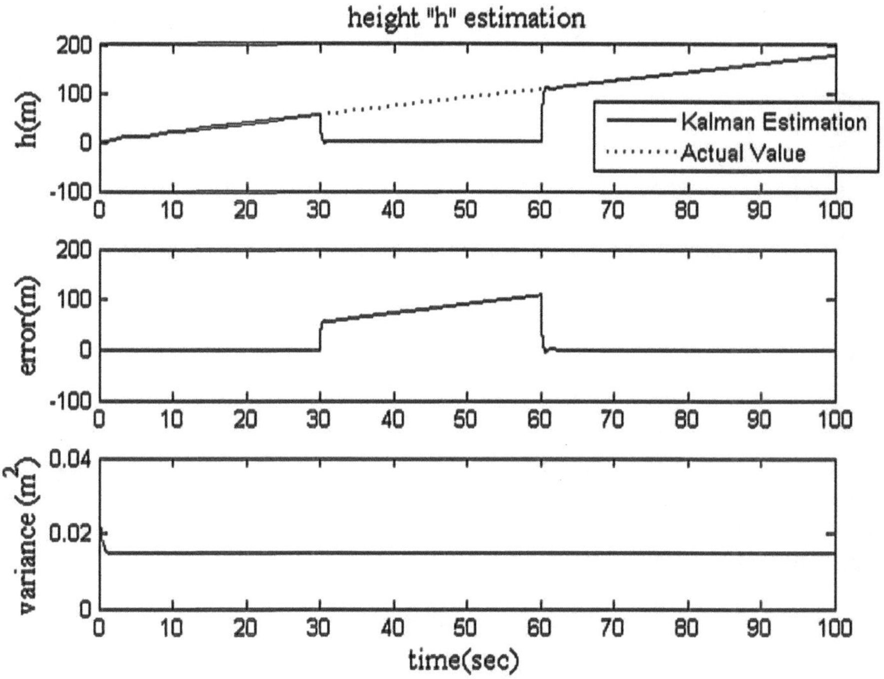

Fig. 5.10 Estimation result for height h by the regular OKF in case of fault of zero output

for the stability analysis of the RKFs. From [21], it is known that the following characteristic polynomial of the system can be used for stability analyses of a linear Kalman filter:

$$\text{Characteristic polynomial} = [zI - \{F_k - K_k H_k F_k\}]\,. \tag{5.14}$$

where z denotes the usual z-transform variable.

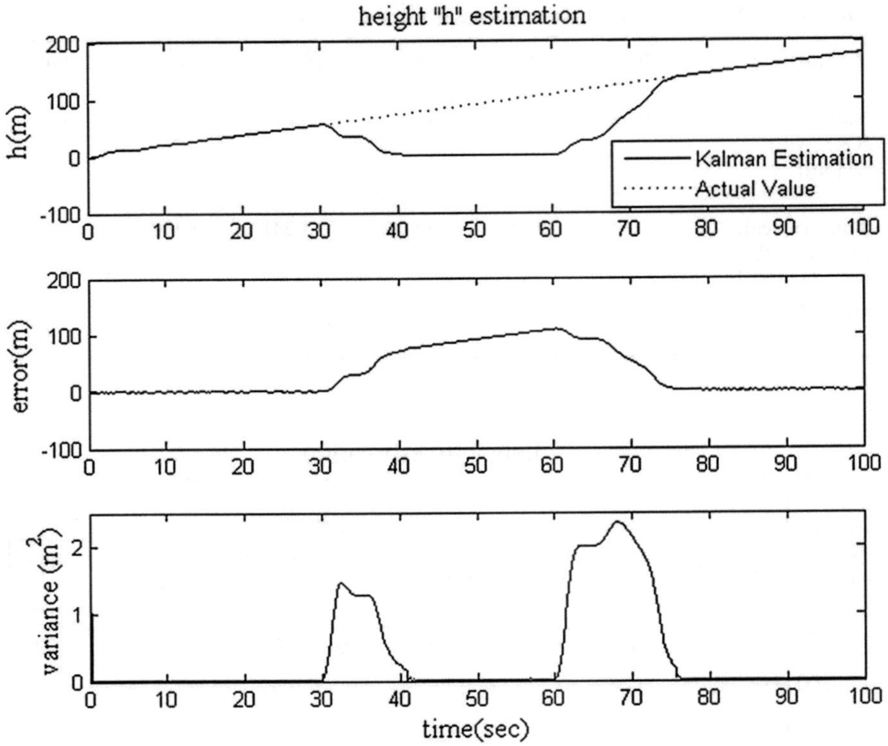

Fig. 5.11 Estimation result for height h by the RKF with SMNSF in case of fault of zero output

The roots of this polynomial provide information about the filter stability. If all the roots lie inside the unit circle in the z-plane, the filter is stable; conversely, if any root lies on or outside the unit circle, the filter is unstable. As a matter of terminology, the roots of the characteristic polynomial are the same as the eigenvalues of:

$$[\{F_k - K_k H_k F_k\}] \tag{5.15}$$

Nonetheless, we assume that a similar investigation is achieved for the RKFs. Let's examine the case for the RKF with MMNSF. Hence, if we substitute the necessary equality instead of Kalman filter gain, Eq. (5.15) becomes:

For the RKF with SMNSF:

$$\left\{F_k - \left[P_{k/k-1} H_k^{\mathrm{T}} \left(H_k P_{k/k-1} H_k^{\mathrm{T}} + S_k R_k\right)^{-1}\right] H_k F_k\right\} \tag{5.16}$$

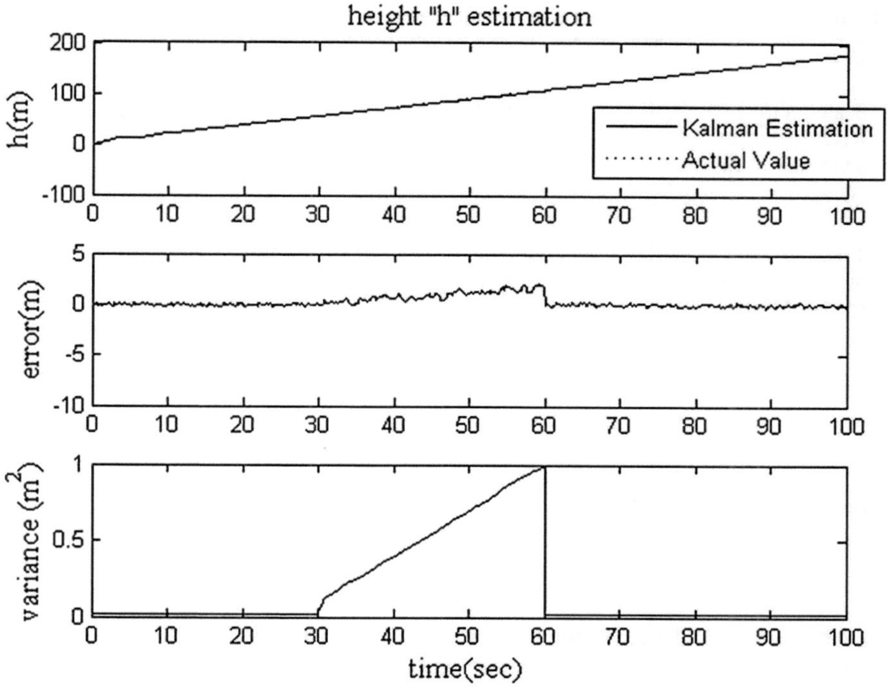

Fig. 5.12 Estimation result for height *h* by the RKF with MMNSF in case of fault of zero output

Table 5.7 Comparison of the mean absolute estimation errors of the regular OKF, RKF with SMNSF, and RKF with MMNSF in case of fault of zero output (for 30 s)

Parameter	Absolute values of error for regular OKF		Absolute values of error for RKF with SMNSF		Absolute values of error for RKF with MMNSF	
	Fault	No fault	Fault	No fault	Fault	No fault
Δu(m/s)	3.45	0.1313	3.129	1.6054	0.1677	0.096
Δw(m/s)	1.1045	0.0709	0.744	0.3416	0.0621	0.0606
Δq(rad/s)	0.1388	0.048	0.3018	0.2252	0.0554	0.0529
$\Delta \theta$(rad)	0.2646	0.0297	0.3057	0.2664	0.0357	0.0278
Δh(m)	80.984	0.0852	71.2394	17.6275	0.9192	0.0954
$\Delta \beta$(rad)	0.0223	0.0236	0.0306	0.019	0.0211	0.0246
Δp(rad/s)	0.0317	0.0336	0.0245	0.0204	0.0316	0.0321
Δr(rad/s)	0.0235	0.0215	0.0273	0.0214	0.0187	0.0221
$\Delta \phi$(rad)	0.0413	0.0384	0.041	0.0331	0.0404	0.041

Table 5.8 MNSF investigation in case of fault of zero output

Parameter	20 s	40 s	45 s	50 s	60 s	90 s
Scale factor of RKF with SMNSF S_k	1	36.0551	1	1	1	1
Trace of scale matrix of RKF with MMNSF $\mathrm{trace}(S_k^*)$	1	125,149	158,130	192,833	275,062	1

For the RKF with MMNSF:

$$\left\{F_k - \left[P_{k/k-1}H_k^{\mathrm{T}}\left(H_kP_{k/k-1}H_k^{\mathrm{T}} + S_k^*R_k\right)^{-1}\right]H_kF_k\right\} \tag{5.17}$$

Now we should search for the roots of Eqs. (5.16) and (5.17). The realized simulations for every malfunction case showed that both RKFs are stable in case of a measurement malfunction. Nonetheless, the RKF with MMNSF is shown to be more stable and, usually, only one root reaches near a point of the unit cycle limit. In Fig. 5.13, the pole-zero map of the RKF with MMNSF is given.

The pole-zero map is obtained for the instantaneous abnormal measurement malfunction case at the 20th second, where the malfunction is implemented. It is not very different from that found for the OKF in the same sampling step. During adaptation with MMNSF, related roots (that belong to the states, the innovation channel of which is corrected via the scale matrix) moves through the unit cycle limit but they do not approach the marginal stability border very closely. On the other hand, for the RKF with SMNSF, where the whole innovation vector is disregarded in case of malfunction, all of the roots move through the stability border (see Fig. 5.14) and, usually, one of them takes a nearly marginal ($|z| = 1$) value, so the stability characteristics worsen.

The simulation results indicate that the proposed RKFs are not at risk of convergence; namely, they are stable. In order to guarantee the accuracy of simulation outputs, some even harsher malfunction cases are tested (far greater malfunctions are implemented). It is seen that the magnitude of the malfunction is not a factor which may break down the stability of the filter and the roots of the RKFs stay inside the unit cycle for every condition.

Moreover, the simulation results indicated the better stability characteristics of the RKF with MMNSF compared to the RKF with SMNSF. For every malfunction case, the roots of the RKF with MMNSF do not approach the stability limit (unit cycle) as much those of the RKF with SMNSF.

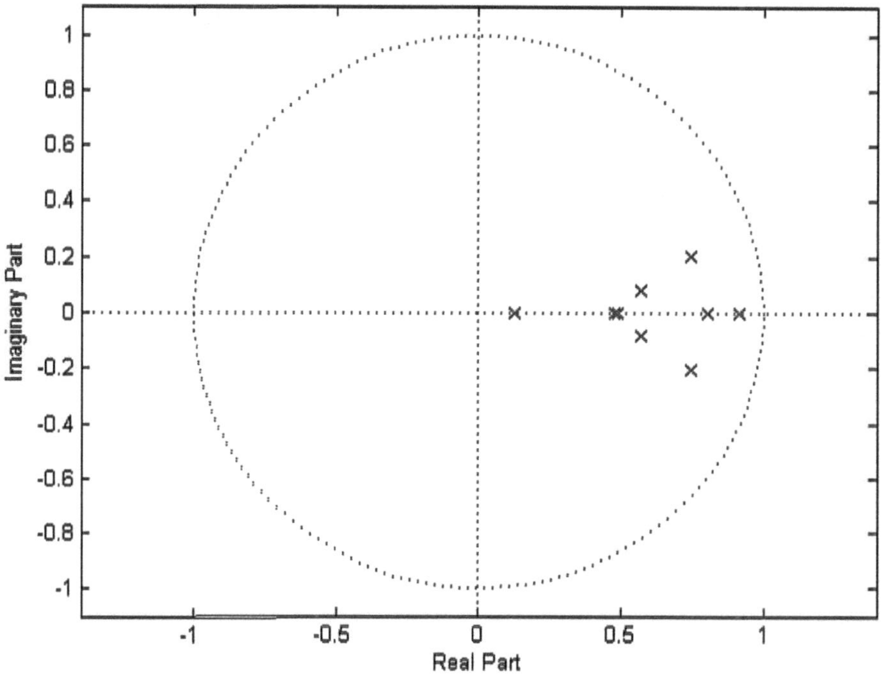

Fig. 5.13 Pole-zero map of the RKF with MMNSF for the case of instantaneous abnormal measurement (at the 20th second)

5.6 Conclusion and Discussion

In this chapter, RKF algorithms with filter gain correction for the case of measurement malfunctions at the current estimation step have been developed. By the use of defined variables named scale factors, faulty measurements are taken into consideration with small weight and the estimations are corrected without affecting the characteristics of the accurate measurements. In the presented RKFs, the filter gain correction is performed only in the case of malfunctions in the measurement system.

RKF algorithms with single and multiple MNSF are proposed. In the first case, the filter is adapted by using single scale factor as a corrective term on the filter gain, and, in the second, a scale matrix composed of multiple factors is used to fix the relevant terms of the Kalman gain matrix individually.

The proposed RKF algorithms with single and multiple MNSF are applied for the state estimation process of a UAV platform. Algorithms are tested for four different measurement malfunction scenarios and the results are compared with the outputs of the OKF for the same cases: instantaneous abnormal measurements, continuous bias of measurements, measurement noise increment, and fault of zero output. In all

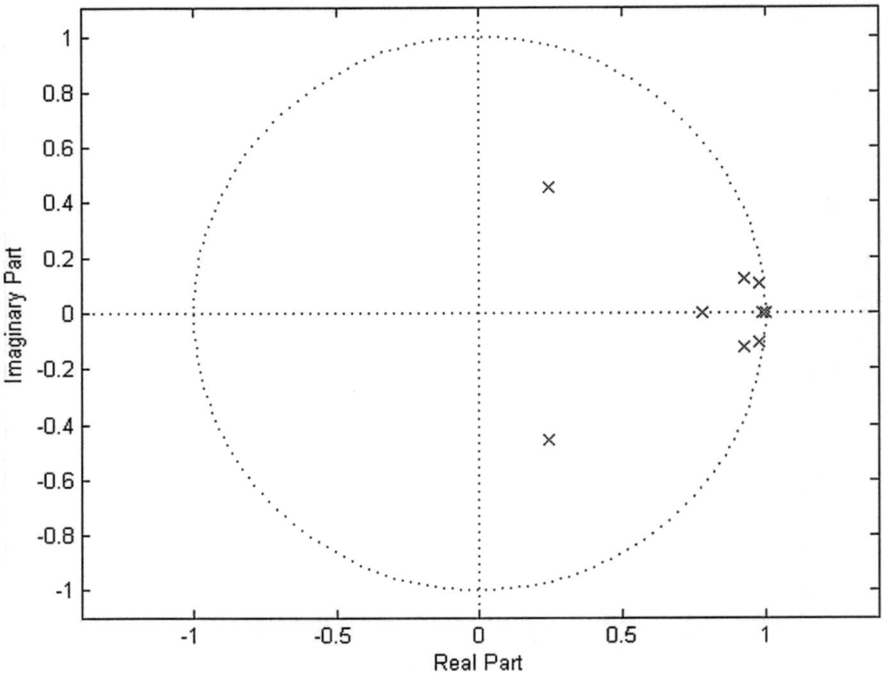

Fig. 5.14 Pole-zero map of the RKF with SMNSF for the case of instantaneous abnormal measurement (at the 20th second)

four scenarios, the OKF becomes faulty while the introduced RKF algorithms stand robust to the measurement faults. Comparison of the simulation results show that the performance of the RKF with MMNSF is better than the RKF with SMNSF. The advantage of the RKF with MMNSF can be especially seen in cases of continuous bias of measurements and fault of zero sensor output. Nonetheless, simulations for the stability analyses indicated that both RKF algorithms are stable in case of a malfunction, but the stability characteristics of the RKF with MMNSF are better than the RKF with SMNSF.

The proposed approach does not require a priori statistical characteristics of the faults and can be used for both linear and nonlinear systems. Furthermore, the presented RKF algorithms are easily implemented in practice and their computational burden is not heavy. These characteristics make the introduced RKF algorithms extremely important from the point of view of supplying reliable parameter estimations for the flight control system of an autonomous UAV.

References

1. White NA, Maybeck PS, DeVilbiss SL (1998) Detection of interference/jamming and spoofing in a DGPS-aided inertial system. IEEE Trans Aerosp Electron Syst 34(4):1208–1217. doi:10.1109/7.722708
2. Zhang Y, Li XR (1997) Detection and diagnosis of sensor and actuator failures using interacting multiple-model estimator. In: Proceedings of the 36th IEEE conference on decision and control, San Diego, CA, December 1997, vol 5, pp 4475–4480. doi:10.1109/CDC.1997.649671
3. Maybeck PS (1999) Multiple model adaptive algorithms for detecting and compensating sensor and actuator/surface failures in aircraft flight control systems. Int J Robust Nonlinear Control 9(14):1051–1070
4. Mehra RK (1970) On the identification of variances and adaptive Kalman filtering. IEEE Trans Autom Control 15(2):175–184
5. Maybeck PS (1982) Stochastic models, estimation, and control, vol I and II. Academic, New York
6. Salychez OS (1994) Special studies in dynamic estimation procedures with case studies in inertial surveying, ENGO 699.26 lecture notes. Department of Geomatics Engineering, University of Calgary, Calgary
7. Mohamed AH, Schwarz KP (1999) Adaptive Kalman filtering for INS/GPS. J Geod 73(4):193–203
8. Wang J, Stewart MP, Tsakiri M (2000) Adaptive Kalman filtering for integration of GPS with GLONASS and INS. In: Proceedings of Geodesy Beyond 2000: the challenges of the first decade, Birmingham, July 1999, pp 325–330
9. Sasiadek JZ, Wang Q (1999) Sensor fusion based on fuzzy Kalman filtering for autonomous robot vehicle. In: Proceedings of the 1999 IEEE international conference on robotics and automation, Detroit, MI, May 1999, vol 4, pp 2970–2975. doi:10.1109/ROBOT.1999.774048
10. Zhang ST, Wei XY (2003) Fuzzy adaptive Kalman filtering for DR/GPS. In: Proceedings of the second international conference on machine learning and cybernetics, Xi-an, China, November 2003, vol 5, pp 2634–2637
11. Escamilla-Ambrosio PJ, Mort N (2003) Hybrid Kalman filter-fuzzy logic adaptive multisensor data fusion architectures. In: Proceedings of the 42nd IEEE conference on decision and control, Maui, HI, December 2003, vol 5, 5215–5220. doi:10.1109/CDC.2003.1272465
12. Hu C, Chen W, Chen Y, Liu D (2003) Adaptive Kalman filtering for vehicle navigation. J Glob Position Syst 2(1):42–47
13. Hide C, Moore T, Smith M (2004) Adaptive Kalman filtering algorithms for integrating GPS and low cost INS. In: Proceedings of the position location and navigation symposium (PLANS), Monterey, CA, April 2004, pp 227–233
14. Ding W, Wang J, Rizos C, Kinlyside D (2007) Improving adaptive Kalman estimation in GPS/INS integration. J Navig 60(3):517–529. doi:10.1017/S0373463307004316
15. Jwo DJ, Weng TP (2008) An adaptive sensor fusion method with applications in integrated navigation. J Navig 61(4):705–721. doi:10.1017/S0373463308004827
16. Kim KH, Lee JG, Park CG (2006) Adaptive two-stage Kalman filter in the presence of unknown random bias. Int J Adapt Contr Signal Process 20(7):305–319. doi:10.1002/acs.900
17. Hajiyev C (2007) Adaptive filtration algorithm with the filter gain correction applied to integrated INS/radar altimeter. Proc Inst Mech Eng Part G J Aerosp Eng 221(5):847–885. doi:10.1243/09544100JAERO173
18. Geng Y, Wang J (2008) Adaptive estimation of multiple fading factors in Kalman filter for navigation applications. GPS Solutions 12(4):273–279. doi:10.1007/s10291-007-0084-6
19. Hajiyev Ch, Soken HE (2009) Adaptive Kalman filter with the filter gain correction applied to UAV flight dynamics. In: Proceedings of the 17th Mediterranean conference on control and automation, Thessaloniki, Greece, 24–26 June 2009

20. Soken HE, Hajiyev C (2010) An adaptive sensor fusion method applied to UAV dynamics. In: Proceedings of the international unmanned vehicles workshop (UVW2010), Istanbul, Turkey, 10–12 June 2010, pp 57–62
21. Hajiyev C, Caliskan F (2003) Fault diagnosis and reconfiguration in flight control systems. Kluwer Academic Publishers, Boston
22. Hajiyev Ch, Soken HE (2009) Adaptive Kalman filter with multiple fading factors for UAV state estimation. In: Proceedings of the 7th IFAC international symposium on fault detection, supervision and safety of technical processes (SafeProcess 2009), Barcelona, Spain, 30 June–3 July, pp 77–82
23. Hajiyev C, Soken HE (2010) Robust estimation of UAV dynamics in the presence of measurement faults. J Aerosp Eng 25(1):80–89

Chapter 6
Estimation of Unmanned Aerial Vehicle Dynamics in the Presence of Sensor/Actuator Faults

6.1 Introduction

An approach for designing reliable systems is to use high-quality components. However, using high-quality components alone does not always reduce the likelihood of system failures, and additional methods must be provided to tolerate faults in the system.

A flight control system is essentially a multi-input/output feedback control system with several nested feedback loops which are designed to meet different performance specifications related to stability augmentation, automatic guidance and navigation, and thrust management. The failure detection and isolation algorithms used in flight computers are implemented with digital computers and are, basically, signal-processing techniques. The reliability of the computer and physical components is very important for the operation of the actual system. A failure may occur in computer subsystems, lanes, sensors, control surfaces, and/or actuators. Since we are examining the failure detection problem from the control engineers' point of view, it is assumed that the failures do not occur in computing subsystems and lanes, but in actuators, control surfaces, and sensors.

In [1], an approach is presented to detect and isolate an aircraft's sensor/actuator faults affecting the mean of the Kalman filter innovation. The effects of the sensor and actuator faults in the innovation are investigated and a decision approach to isolate the fault is proposed. Although it proposes to be an effective method for fault isolation, this work examines the sensor and actuator faults affecting only the mean of the innovation, so the method works under some certain conditions and it is not appropriate for applying in case of various different fault types. Nonetheless, the given measurement noise covariance adaptation procedure isdifficult to realize.

C. Hajiyev et al., *State Estimation and Control for Low-cost Unmanned Aerial Vehicles*, DOI 10.1007/978-3-319-16417-5_6

Another approach is to scale the noisec ovariance matrix by multiplying it by a time-dependent variable. An adaptive Kalman filter is used to tune the measurement and process noise covariance matrices R and Q, respectively. The philosophy of estimating (or scaling) the Q and R matrices follows one of these scenarios [2]:

- Fixing Q and varying R by trial and error until the realistic values are found, such that the filter gives stable state estimates. In this case, the Q matrix should be completely known.
- Varying the Q matrix if R is completely known and fixed to a certain value.
- Varying Q and R simultaneously; in this case, neither is known.

When the process noise covariance matrix is scaled (Q-adaptation), this means that there is an actuator fault to be compensated. Otherwise, when the measurement noise covariance matrix is scaled (R-adaptation), it is a sensor fault, which is dealt with. In [3, 4], an adaptation procedure based on the single fading factor and in [5], a scheme based on multiple fading factors are presented. The essence of using multiple fading factors is the different effect of the fault on the estimation performance of each estimated state. Especially for complex multivariable systems, the effect of the fault on each state should be investigated carefully and, instead of a single factor, a matrix formed of multiple fading factors (such that it weights the adaptation differently for each state) should be used. Nonetheless, as a disadvantage, these papers [3–5] take only the Q-adaptation procedure into consideration and do not examine the R-adaptation methods. Per contra, the R-adaptation is investigated for possible sensor faults in [6–9] with a considerably simpler technique, but the Q-adaptation is not investigated. In these studies, it is shown that, if there is a malfunction in the measurement system, the robust adaptive Kalman filter (RAKF) algorithm can be utilized and, by the use of a measurement noise scale factor (MNSF) as a multiplier of the measurement noise covariance, insensitiveness of the filter to the current sensor faults can be ensured. In [10], a similar adaptation algorithm is applied to the unscented Kalman filter for the nonlinear estimation problem, just as for the R-adaptation.

Moreover, there are also published works in the literature which are interested in both the R- and Q-adaptations [11–13]. However, in these papers, an isolation scheme for the fault is absent; it is not described how to operate an RAFK in the presence of sensor/actuator faults. These filters can detect whether there is a fault in the system or not, but the type of fault cannot be determined. Instead of using a fault isolation scheme, a simultaneous adaptation procedure for the R- and Q-adaptations is performed in these studies. However, as indicated in [2], full estimation of R and Q based on covariance matching is questionable, since these two values rely on each other in the covariance matching equation. Consequently, these filters cannot estimate the unmanned aerial vehicle (UAV) dynamics reliably in the presence of sensor/actuator faults. This problem can be solved through the combined fault isolation and robust adaptive filtration techniques. Besides, the methods proposed in these papers are all based on the single adaptive factor and, for complex multivariable systems, they may not be sufficiently accurate.

In this chapter, an RAKF is developed that incorporates the R- and Q-adaptation procedures and adapts itself against sensor/actuator faults in order to ensure robustness. The filter first detects the type of fault (either sensor or actuator) and then isolates it. After that, the appropriate adaptation (R or Q) is applied such that the estimation characteristic is not deteriorated. Unlike the existing studies, the proposed adaptation methods for both the R- and Q-adaptations are simplistic and, with a small modification, can be applied easily to the Kalman filter. Also in this chapter, it is shown how to incorporate these two different adaptation processes such that the fault is isolated and, so, the RAKF algorithm is a complete estimator including the fault detection and isolation processes for the case of sensor/actuator faults. The RAKF is applied for the state estimation procedure of a UAV and its effectiveness is investigated by comparisons with the optimal Kalman filter (OKF) in case of faults.

6.2 Q-adaptation Using Multiple Adaptive Factors

When there is an actuator fault in the system, it results in changes in the control distribution matrix. In a similar manner to the R-adaptation procedure, in case of an actuator fault, the real error of the innovation covariance will exceed the theoretical covariance. Thus, the basic premise of the Q-adaptation is also to obtain an appropriate multiplier matrix for the Q matrix such that the real and theoretical values of the innovation covariance match. Hence, if a fading matrix (Λ_k) composed of multiple fading factors is added into the algorithm as [14–17]:

$$\frac{1}{\mu}\sum_{j=k-\mu+1}^{k} \tilde{e}_j\tilde{e}_j^{\mathrm{T}} = H_k\left(F_kP_{k-1/k-1}F_k^{\mathrm{T}} + \Lambda_kG_kQ_kG_k^{\mathrm{T}}\right)H_k^{\mathrm{T}} + R_k \tag{6.1}$$

then the fading matrix can be determined as:

$$\Lambda_k = H_k^{-1}\left(\frac{1}{\mu}\sum_{j=k-\mu+1}^{k} \tilde{e}_j\tilde{e}_j^{\mathrm{T}} - H_kF_kP_{k-1/k-1}F_k{}^{\mathrm{T}}H_k^{\mathrm{T}} - R_k\right)\left(G_kQ_kG_k{}^{\mathrm{T}}H_k^{\mathrm{T}}\right)^{-1} \tag{6.2}$$

For a specific case where all the states are measured as ($H_k = I$), as in the case examined here, Eq. (6.2) reduces to:

$$\Lambda_k = \left(\frac{1}{\mu}\sum_{j=k-\mu+1}^{k} \tilde{e}_j\tilde{e}_j^{\mathrm{T}} - F_kP_{k-1/k-1}F_k{}^{\mathrm{T}} - R_k\right)\left(G_kQ_kG_k{}^{\mathrm{T}}\right)^{-1} \tag{6.3}$$

In a similar manner to the R-adaptation, the obtained fading matrix should be diagonalized, since the Q matrix must be a diagonal, positive-definite matrix:

$$\Lambda^* = \text{diag}\left(\lambda_1^*, \lambda_2^*, \ldots, \lambda_n^*\right), \tag{6.4}$$

where:

$$\lambda_i^* = \max\{1, \Lambda_{ii}\} \qquad i = 1, \ldots, n. \tag{6.5}$$

Here, Λ_{ii} represents the ith diagonal element of the matrix Λ. Apart from that point, if there is an actuator fault in the system, Λ_k^* must be included in the estimation process as:

$$P_{k/k-1} = F_k P_{k-1/k-1} F_k^{\mathrm{T}} + \Lambda_k^* G_k Q_k G_k^{\mathrm{T}} \tag{6.6}$$

6.3 Integration Scheme for the Q- and R-adaptation Methods

Due to the scale or fading matrix, the covariance of the estimation error of the RAKF increases in comparison with the OKF. Therefore, the RAKF algorithm is used only when the there is a fault and, in all other cases, the procedure is run optimally with the regular OKF. The fault detection process is applied by checking a statistical function at each estimation step. At that point, the two following hypotheses may be introduced:

- γ_0; the system is operating normally
- γ_1; there is a malfunction in the estimation system

In order to detect the faults, the statistical function used is defined as:

$$\beta_k = \tilde{e}_k^{\mathrm{T}} \left[H_k P_{k/k-1} H_k^{\mathrm{T}} + R_k\right]^{-1} \tilde{e}_k. \tag{6.7}$$

This statistical function has a χ^2 distribution with M degrees of freedom (M is the dimension of the innovation vector).

If the level of significance, α, is selected as:

$$P\left\{\chi^2 > \chi^2_{\alpha,M}\right\} = \alpha; \quad 0 < \alpha < 1, \tag{6.8}$$

then the threshold value, $\chi^2_{\alpha,M}$, can be found. Hence, when the hypothesis γ_1 is correct, the statistical value of β_k will be greater than the threshold value $\chi^2_{\alpha,M}$, i.e.:

$$\begin{aligned} \gamma_0 &: \beta_k \leq \chi^2_{\alpha,M} \qquad \forall k \\ \gamma_1 &: \beta_k > \chi^2_{\alpha,M} \qquad \exists k. \end{aligned} \tag{6.9}$$

On the other hand, after detecting the fault in the system, the key point is detecting the type of fault (either a sensor or an actuator fault). After that, the appropriate adaptation (R- or Q-adaptation) may be applied [15–17]. The fault isolation can be realized by an algorithm similar to that proposed for aircraft in [18]. When a regular Kalman filter is used, the decision statistics change regardless of whether the fault is in the sensors or in the actuators. If an RKF based on the Doyle–Stein condition is used, it is easy to distinguish between sensor and actuator faults. The Kalman filter that satisfies the Doyle–Stein condition is referred in [18], as the RKF is insensitive to actuator failures and may also be used for fault isolation in our study. After that, the appropriate adaptation (R- or Q-adaptation) may be applied. The Doyle–Stein condition is as follows:

$$K(I + H\phi_s K)^{-1} = B_s(H\phi_s B_s)^{-1}. \tag{6.10}$$

Here, K is the filter gain, I is a unit matrix, H is the system measurement matrix, B_s is the control distribution matrix in continuous time, and $\phi_s = (sI - F_s)^{-1}$, where F_s is the system matrix in continuous time.

The RKF is very useful for isolating sensor and control surface failures, as it is insensitive to the latter failures. If the Kalman filter process noise covariance matrix is chosen as

$$Q(q_r) = Q + q_r^2 B_s V B_s^{\mathrm{T}} \tag{6.11}$$

then the filter becomes robust against the actuator faults. Here, Q is the process noise covariance matrix for the nominal plant, q_r is a parameter that approaches to infinity as the filter's robustness increases, and V is a positive-definite symmetric matrix.

If a sensor fault occurs, the R-adaptation is realized; else, the procedure is continued with the Q-adaptation. The algorithm architecture given in Fig. 6.1 summarizes the integrated R- and Q-adaptation procedures. This new filter, where the robust and adaptive Kalman filters are integrated, can be called the RAKF.

6.4 Numerical Example

In order to examine the effectiveness of the proposed RAKF, it was applied for the state estimation of the Zagi UAV. The simulations are realized in 1,000 steps for a period of 100 s with 0.1 s of sampling time, Δt.

In case of sensor/actuator fault, the simulations are also carried out for the OKF so as to compare the results with the RAKF algorithm and understand better the efficiency of the proposed algorithm.

Nonetheless, for the fault detection process, $\chi^2_{\alpha,M}$ is taken as 21.7, and this value comes from the Chi-square distribution when the number of degrees of freedom is 9 and the reliability level is 99 %.

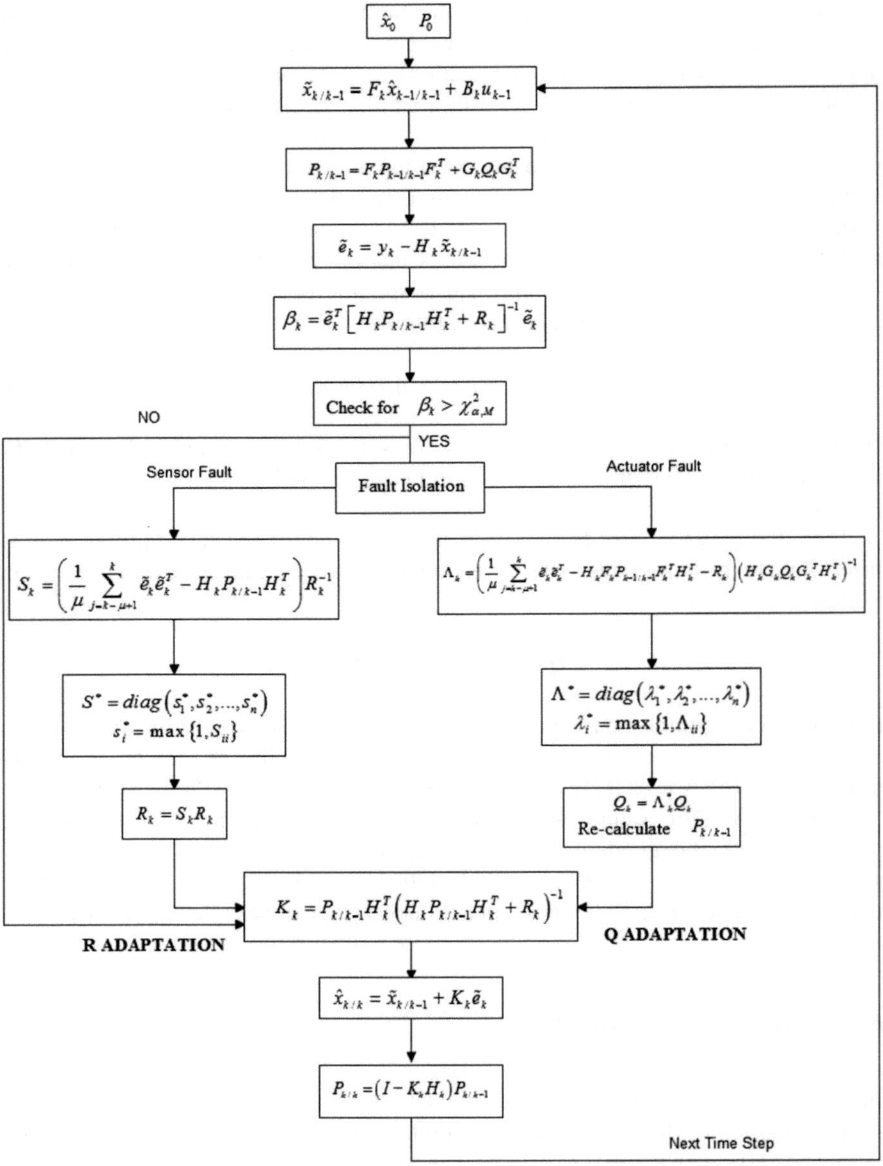

Fig. 6.1 System architecture for the robust adaptive Kalman filter (RAKF) algorithm

All the figures given in this section consist of three subfigures. The top figure gives the OKF or RAKF state estimation results and the actual values for comparison. The middle figure shows the error of the estimation process based on the actual state values of the UAV. The bottom figure indicates the variance of the estimation that is obtained as the diagonal elements of the Kalman filter estimation error covariance matrix.

During the simulations, for testing the fault detection and isolation procedures and also the RAKF algorithm, two different faults are implemented to the systems. Firstly, a sensor fault is formed by adding a constant term to the measurement of the pitch angle (the constant term is nearly of the same magnitude as the measurement, so the measurement's magnitude is doubled for those instants of time), $\Delta\theta$, in between the 30th and 50th seconds. Furthermore, in order to check that fault isolation can be achieved successfully and switching from one adaptation stage to the other is realized properly, as another fault, the actuator fault is simulated by taking the first column elements of the longitudinal control distribution matrix to be nearly zero in between the 80th and 90th seconds. The graph for the statistical values of β_k in case of using the OKF is shown in Fig. 6.2. As can be seen, except for the periods where the sensor and actuator failures occur, β_k is lower than the threshold value.

On the contrary, when one of these failures occurs, β_k grows rapidly and exceeds the threshold value. Hence, the γ_1 hypothesis is judged to be true and it is shown that, by using such a procedure, the fault can be detected. The results for the fault isolation process are given in Fig. 6.3.

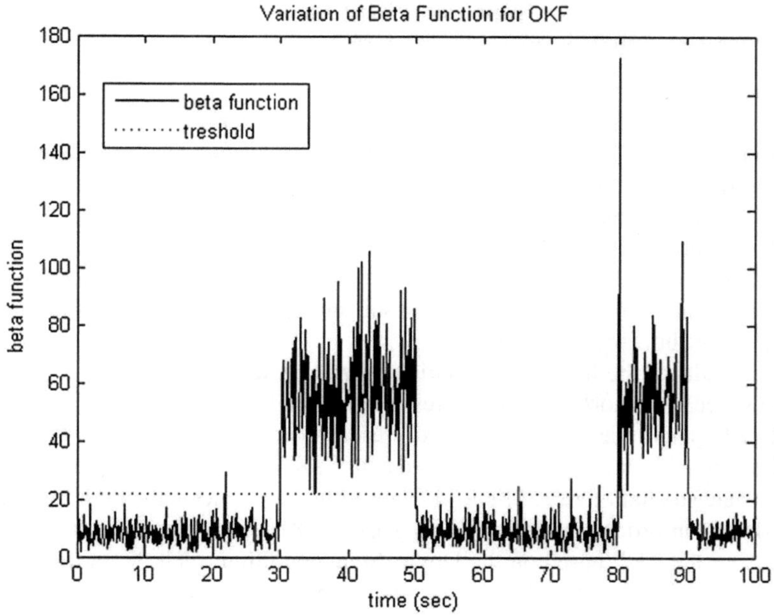

Fig. 6.2 Behavior of the statistics β_k in case of measurement failure (in between the 30th and 50th seconds) and actuator failure (in between the 80th and 90th seconds) when the optimal Kalman filter (OKF) is used

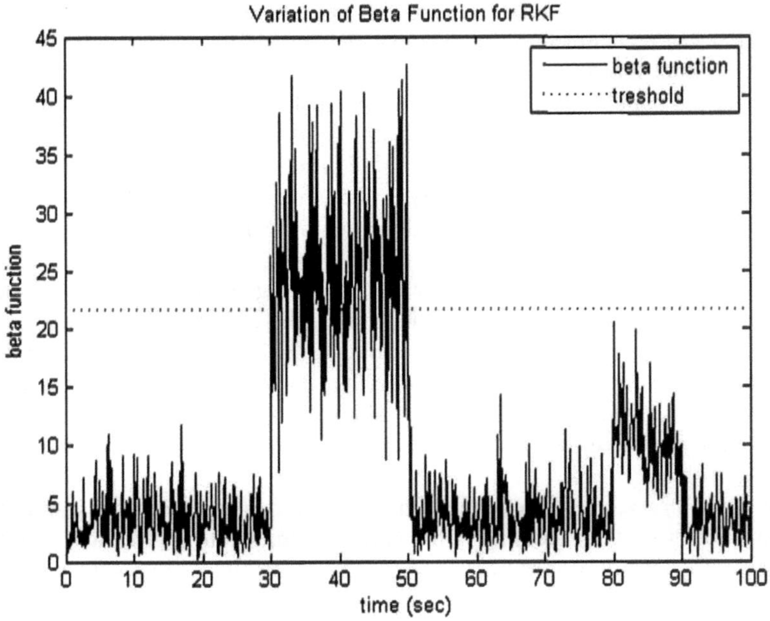

Fig. 6.3 Behavior of the statistics β_k in case of measurement failure (in between the 30th and 50th seconds) and actuator failure (in between the 80th and 90th seconds) when the robust Kalman filter (RKF), which is insensitive to actuator failures, is used

As mentioned, the RKF, which is insensitive to actuator failures, is used for this isolation process. The behavior of the statistics β_k in the case of sensor/actuator failures shows that detecting the actuator failure is not possible when the RKF is used, since it is insensitive against such faults; on the other hand, the sensor failure is detected immediately. Hence, the fault isolation can be realized by using such a procedure. The RKF, which is insensitive to actuator failures and sensitive to sensor failures, is employed to isolate the sensor and actuator failures in the proposed RAKF algorithm (see Fig. 6.1).

The results of the first fault scenario realized for testing the RAKF algorithm in the presence of sensor/actuator failures are given in Figs. 6.4 and 6.5. It is clear that the RAKF gives superior estimation results compared to the OKF in both cases.

The type of fault is precisely detected and the RAKF tunes itself according to the appropriate adaptation rule. The RAKF algorithm first detects the fault by using the fault detection process given with Eqs. (6.7), (6.8), and (6.9). Then, it isolates the fault using the RKF based on the Doyle–Stein condition, such that it distinguishes the sensor and actuator faults. As the final step, it adapts the R matrix in between the 30th and 50th seconds and the Q matrix in between the 80th and 90th seconds.

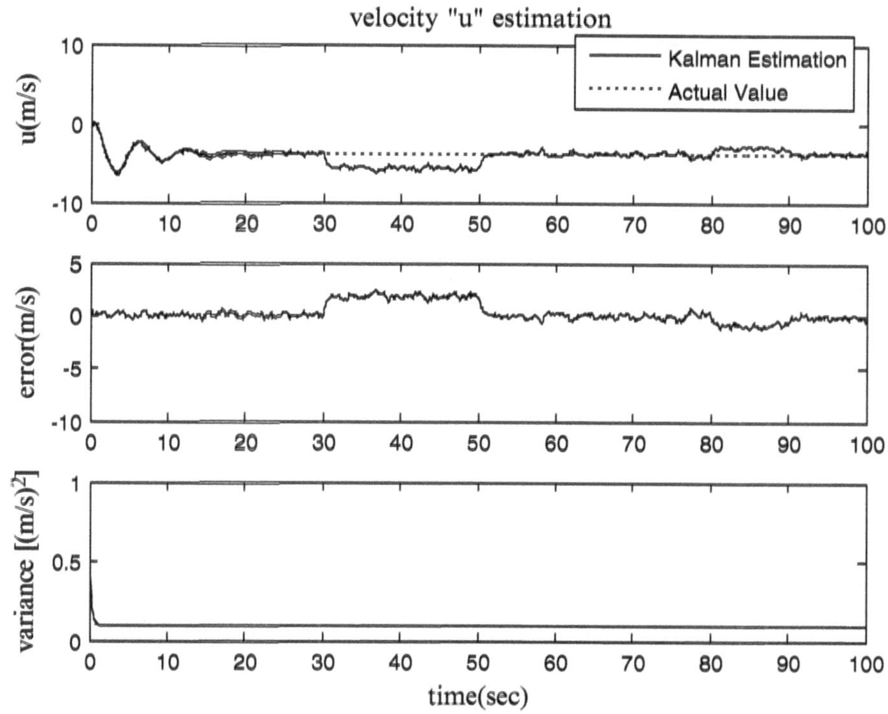

Fig. 6.4 *u* velocity estimation via the OKF in case of actuator and bias sensor faults

As seen from the figures, it ensures accurate estimation results for the whole process, while the OKF fails in both the sensor fault and the actuator fault conditions. This proves us three facts:

- The fault detection process is working correctly and the fault in the system is detected without any time delay
- By using the RKF based on the Doyle–Stein condition, the type of fault is distinguished and the RAKF decides to use the right adaptation procedure against the fault
- Both of the proposed adaptation methods are functioning properly

Nevertheless, the performance of both filters, the OKF and the RAKF, in case of fault can be understood more clearly by examining Table 6.1.

In the table, the mean absolute values of error, which are calculated for 10 s, are given for all three conditions: sensor fault, actuator fault, and no fault. The tabulated results show that the estimations by the OKF worsen for all of the longitudinal states in case of sensor/actuator fault, while the RAKF retains accurate

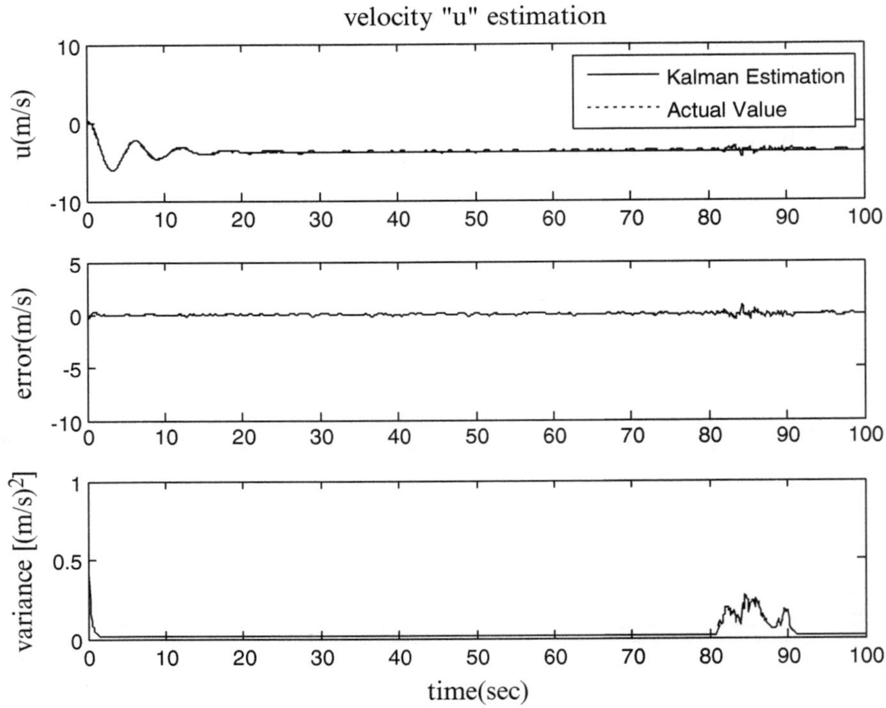

Fig. 6.5 *u* velocity estimation via the RAKF in case of actuator and bias sensor faults

Table 6.1 Comparison of the mean absolute values of error for the optimal Kalman filter (OKF) and the robust adaptive Kalman filter (RAKF) in case of actuator and bias sensor faults

Parameter	Mean absolute value of error for regular OKF			Mean absolute value of error for RAKF		
	Sensor fault	Operating normally	Actuator fault	Sensor fault	Operating normally	Actuator fault
Δu(m/s)	1.8799	0.1891	0.8116	0.0528	0.0578	0.1815
Δw(m/s)	0.6762	0.0931	1.1939	0.0649	0.0611	0.2678
Δq(rad/s)	0.0604	0.0621	0.1602	0.0521	0.0542	0.0686
$\Delta \theta$(rad)	0.6497	0.053	0.0796	0.0161	0.0132	0.0325
Δh(m)	0.5493	0.1059	0.1273	0.0644	0.0636	0.0932
$\Delta \beta$(rad)	0.0319	0.0302	0.0338	0.0123	0.0155	0.0165
Δp(rad/s)	0.0514	0.0556	0.0538	0.0282	0.0257	0.0304
Δr(rad/s)	0.0508	0.045	0.0465	0.0115	0.0146	0.0163
$\Delta \phi$(rad)	0.0758	0.0786	0.0805	0.0167	0.0192	0.0349

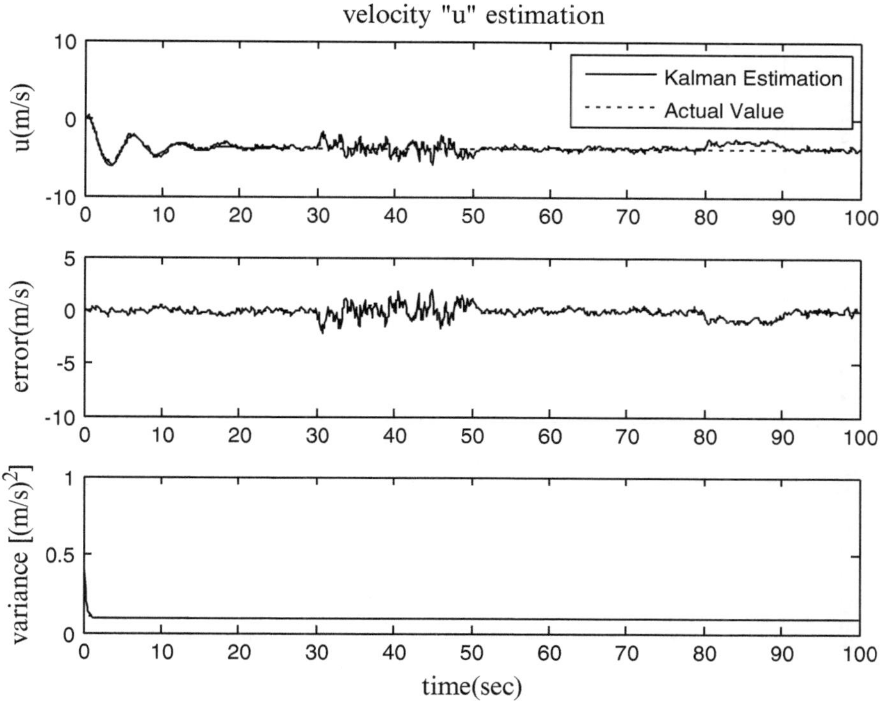

Fig. 6.6 *u* velocity estimation via the OKF in case of actuator and noise increment sensor faults

estimation behavior for the whole process. Because of the decoupled dynamics used for the UAV, any fault in the longitudinal channel (e.g., sensor fault for the sensor measuring $\Delta\theta$ or the actuator fault affecting the longitudinal control derivatives) does not have an effect on the lateral states as expected. This fact makes the importance of using multiple adaptive factors for both adaptation procedures more distinct.

As the second testing scenario for the RAKF algorithm, this time, the sensor fault is characterized by multiplying the variance of the noise of the measurement for the Δu velocity component with a constant term in between the 30th and 50th seconds, while the actuator fault is the same as the first scenario. As Figs. 6.6 and 6.7 show, again, the OKF outputs involve error, while the RAKF algorithm achieves estimation of the states accurately. Nonetheless, another consequence of the second fault scenario is the good performance of the fault isolation scheme; regardless of the type of sensor fault, the isolation scheme can distinguish the type of fault effectively and the RAKF works properly.

In order to clarify the results, a comparison of the mean absolute values of error is given in Table 6.2.

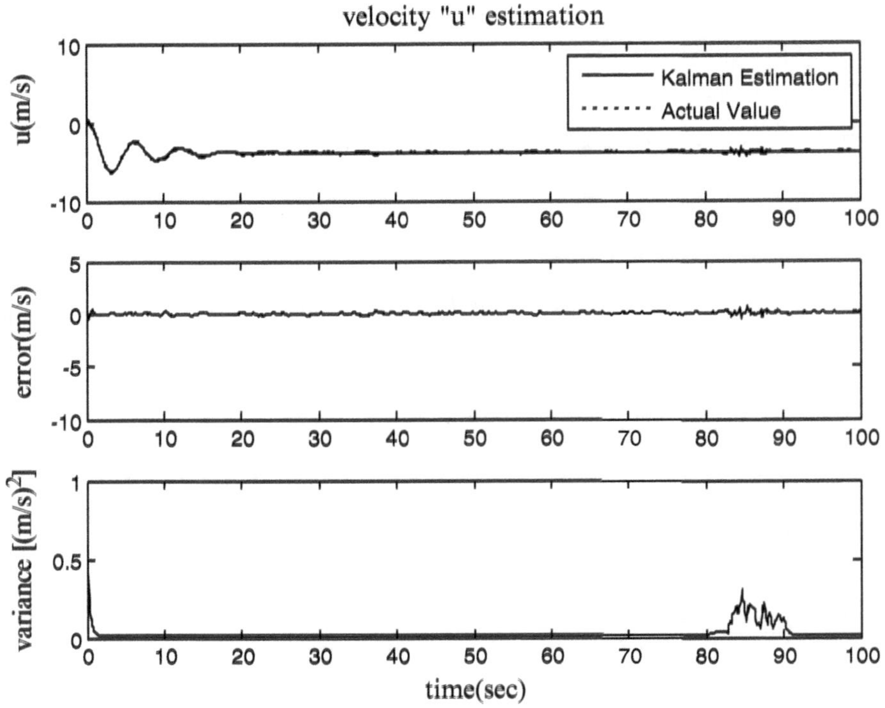

Fig. 6.7 *u* velocity estimation via the RAKF in case of actuator and noise increment sensor faults

Table 6.2 Comparison of the mean absolute values of error for the OKF and RAKF in case of actuator and noise increment sensor faults

Parameter	Mean absolute value of error for regular OKF			Mean absolute value of error for RAKF		
	Sensor fault	Operating normally	Actuator fault	Sensor fault	Operating normally	Actuator fault
Δu(m/s)	0.6235	0.2317	0.8352	0.0819	0.0635	0.1303
Δw(m/s)	0.2588	0.1482	1.1963	0.0687	0.0626	0.2621
Δq(rad/s)	0.0611	0.0673	0.1505	0.055	0.0508	0.0711
$\Delta \theta$(rad)	0.0653	0.0621	0.0688	0.0115	0.0155	0.0376
Δh(m)	0.1101	0.1	0.1323	0.0826	0.0733	0.0785
$\Delta \beta$(rad)	0.0355	0.0392	0.0319	0.0163	0.0149	0.0134
Δp(rad/s)	0.0497	0.0465	0.0553	0.0308	0.026	0.0265
Δr(rad/s)	0.0456	0.0443	0.0449	0.0135	0.013	0.0112
$\Delta \phi$(rad)	0.08	0.0694	0.0695	0.0262	0.0196	0.0359

6.5 Conclusion and Discussion

In this chapter, first, the Q-adaptation with multiple adaptive factors for actuator fault compensation is presented. Then, the RAKF with the R- and Q-adaptations against sensor/actuator failures is proposed.

The proposed RAKF algorithm with the multiple adaptive factors incorporates the R- and Q-adaptation procedures and adapts itself against sensor/actuator faults in order to ensure robustness. The filter first detects the type of fault (either sensor or actuator) and then isolates it. After that, the appropriate adaptation (R- or Q-adaptation) is applied such that the estimation characteristic is not deteriorated. Unlike existing studies, the proposed adaptation methods for both the R- and Q-adaptations are simplistic and can be applied easily with a small modification to the Kalman filter.

In this chapter, it has been shown how to incorporate these two different adaptation processes such that the fault is isolated and, so, the RAKF algorithm is a complete estimator including fault detection and isolation processes for the case of sensor/actuator faults. When a regular Kalman filter is used, the decision statistics change regardless of whether the fault is in the sensors or in the actuators, whereas when an RKF based on the Doyle–Stein condition is used, it is easy to distinguish the sensor and control surface/actuator faults. The RAKF is applied for the state estimation procedure of a UAV and its effectiveness is investigated by comparisons with the OKF in case of faults. The proposed RAKF algorithm may play an important role for UAV control systems since it gives accurate estimation results despite the sensor or actuator faults. Regarding the harsh environment where UAVs are generally used, encountering any kind of fault has a high probability and, so, preferring the proposed RAKF algorithm instead of the regular OKF may bring a significant advantage.

References

1. Hajiyev C, Caliskan F (2000) Sensor/actuator fault diagnosis based on statistical analysis of innovation sequence and robust Kalman filtering. Aerosp Sci Technol 4:415–422. doi:10.1016/S1270-9638(00)00143-7
2. Almagbile A, Wang J, Ding W (2010) Evaluating the performances of adaptive Kalman filter methods in GPS/INS integration. J Glob Position Syst 9:33–40. doi:10.5081/jgps.9.1.33
3. Hu C, Chen W, Chen Y, Liu D (2003) Adaptive Kalman filtering for vehicle navigation. J Glob Position Syst 2(1):42–47
4. Ding W, Wang J, Rizos C, Kinlyside D (2007) Improving adaptive Kalman estimation in GPS/INS integration. J Navig 60(3):517–529. doi:10.1017/S0373463307004316
5. Geng Y, Wang J (2008) Adaptive estimation of multiple fading factors in Kalman filter for navigation applications. GPS Solutions 12(4):273–279. doi:10.1007/s10291-007-0084-6
6. Hajiyev C (2007) Adaptive filtration algorithm with the filter gain correction applied to integrated INS/radar altimeter. Proc Inst Mech Eng Part G J Aerosp Eng 221(5):847–855. doi:10.1243/09544100JAERO173

7. Hajiyev Ch, Soken HE (2009) Adaptive Kalman filter with the filter gain correction applied to UAV flight dynamics. In: Proceedings of the 17th Mediterranean conference on control and automation, Thessaloniki, Greece, 24–26 June 2009
8. Hajiyev Ch, Soken HE (2009) Adaptive Kalman filter with multiple fading factors for UAV state estimation. In: Proceedings of the 7th IFAC international symposium on fault detection, supervision and safety of technical processes (SafeProcess 2009), Barcelona, Spain, 30 June–3 July, pp 77–82
9. Hajiyev C, Soken HE (2010) Robust estimation of UAV dynamics in the presence of measurement faults. J Aerosp Eng 25(1):80–89
10. Soken HE, Hajiyev C (2010) Pico satellite attitude estimation via robust unscented Kalman filter in the presence of measurement faults. ISA Trans 49(3):249–256. doi:10.1016/j.isatra.2010.04.001
11. Jwo DJ, Weng TP (2008) An adaptive sensor fusion method with applications in integrated navigation. J Navig 61(4):705–721. doi:10.1017/S0373463308004827
12. Jwo DJ, Chang SC (2009) Particle swarm optimization for GPS navigation Kalman filter adaptation. Aircr Eng Aerosp Technol Int J 81:343–352. doi:10.1108/00022660910967336
13. Kim KH, Lee JG, Park CG (2006) Adaptive two-stage Kalman filter in the presence of unknown random bias. Int J Adapt Control Signal Process 20(7):305–319. doi:10.1002/acs.900
14. Hajiyev Ch, Soken HE (2011) Fault tolerant estimation of UAV dynamics via robust adaptive Kalman filter. In: Proceedings of the special international conference on complex systems: synergy of control, communications and computing (COSY 2011), Ohridi, Republic of Macedonia, 16–20 September 2011, pp 311–320
15. Hajiyev C, Soken HE (2013) Robust adaptive Kalman filter for estimation of UAV dynamics in the presence of sensor/actuator faults. Aerosp Sci Technol 28(1):376–383. doi:10.1016/j.ast.2012.12.003
16. Hajiyev C, Soken HE (2013) Robust self-adaptive Kalman filter with the R and Q adaptations against sensor/actuator failures. In: Zhang WJ (ed) Self-organization: theories and methods. Nova Science Publishers, New York
17. Hajiyev C, Soken HE (2015) Fault tolerant estimation of UAV dynamics via robust adaptive Kalman filter. In: Dimirovski GM (ed) Complex systems: relationships between control, communications and computing. Springer (in press), Berlin
18. Hajiyev C, Caliskan F (2005) Sensor and control surface/actuator failure detection and isolation applied to F-16 flight dynamics. Aircr Eng Aerosp Technol Int J 77:152–160

Chapter 7
Fault Detection, Isolation, and Data Fusion for Unmanned Aerial Vehicle Air Data Systems

7.1 Introduction

For unmanned aerial vehicles (UAVs) to carry out their assigned tasks and complete missions successfully, appropriate air measurement data should be derived in case of fault occurrence in sensor measurements. In order to produce appropriate data, the following should be done: detecting and isolating the faults in the measurement system, determining the most appropriate data, using other measurement data for estimation purposes, and deriving the most appropriate data using the available measurements (decision algorithm). There have been many studies using different techniques investigating the subject [1–10]. Some of them are given below.

In [1], a method using the extended Kalman filtering (EKF) structure is proposed and the efficacy of the method is shown using a space vehicle dynamic model which includes redundant sensors to measure the parameters. Although using multiple extended Kalman filters together in nonlinear systems for fault detection is a known method, generally, the initial value of the covariance matrix is chosen by expertise; however, in this study, a new method for determining the optimal covariance matrix is proposed.

Two different fusion methods for a UAV that integrate inertial navigation system (INS) and accurate dynamic model measurements are proposed in [2]. In order to reduce the growing errors of the INS and obtain better measurements, the measurements of a dynamic model are integrated using EKF. In that study, the errors between the INS and flight vehicle model (FVM) measurements are estimated with the filtering technique using the error model of the system. The first proposed method is based on comparing speed and Euler angle measurements produced by the INS and the FVM. In this first method, inertial measurement unit (IMU) acceleration and angular speed errors, INS speed and Euler angle errors, and FVM speed and Euler angle errors are determined and used to obtain better measurement data. In the second method, the estimated values of acceleration and angular speed taken from

C. Hajiyev et al., *State Estimation and Control for Low-cost Unmanned Aerial Vehicles*, DOI 10.1007/978-3-319-16417-5_7

the FVM are used for the calibration of the IMU. The errors in the acceleration and angular speed measurements are estimated using an error model of the FVM.

In [3], a sensor fault detection method using data fusion techniques is investigated. In this study, the EKF method is used to integrate data. An adaptive method which updates the sensor noise covariance matrix according to residual values covariance is used in EKF. This new method is called the adaptive modified extended Kalman filtering (AMEKF) algorithm. The AMEKF system estimates the system noise and sensor noise covariance matrices, which can affect normal filter working conditions, and finds calibration errors.

A weak model-based method for the detection of sensor abrupt faults is considered in [4]. In that study, some characteristics of the sensors are used to derive the necessary signal for fault detection. The method is simulated using a small aircraft and an environment model, which includes sensor noise, wind effect, and turbulence. In the simulation, angle of attack (AOA) sensor faults are considered.

In [5], a new method is developed for a nonlinear flight model to detect input and output sensor faults. In the study, a polynomial residual generator structure is used. This approach, which is robust against faults, can filter the error signal and decrease the effect of the additive noise and error that occurs due to linearization and measurement. The input errors in sensors are separated as known input, disturbance, and faults vectors. The designed system describes each input–output relation by polynomials and uses residuals to detect and isolate faults. Different fault detection methods are also used for the same fault conditions to better evaluate the proposed method. For this purpose, nonlinear geometric approach (NLGA), unknown input Kalman observers (UIKF), and neural networks (NN) are investigated.

In [6], the NN method is used for fault detection and isolation (FDI), and is applied to a nonlinear UAV model. In the study, a radial basis function (RBF) type neural NN is designed for modeling purposes. A novel residual generation technique is also used to overcome wrong fault error situations. A nonlinear UAV model is considered and longitudinal equations are used to simulate the fault situation in the pitch angle gyro.

In [7], a similar system to that used in [4] is investigated. The characteristics of the sensor signal are again used for FDI. The design is the same as that described in [4], except for the dynamic dependency filter structure, which changed from linear autoregressive to pooled autoregressive (PAR). The idea comes from changing the linear structure of the filter in [4] to obtain better results. The results are again positive, as in [4].

In [8], interacting multiple model (IMM) and unscented Kalman filter (UKF) methods are used with nonlinear aircraft dynamics for fault detection purposes. The differences between using the INS equations and a dynamic model are shown for the situations without faults. The IMM-UKF structure is proved to estimate the states of the system in the presence of one faulty sensor. In conclusion, the IMM-UKF structure works properly if used together for fault detection purposes and is an effective new way of finding the faults in the system.

In [9], an FDI system is proposed for a nonlinear model. There are widely used FDI algorithms for linear systems; however, these methods are not sufficient for nonlinear systems and can cause errors in fault detection. For this reason, an input estimation structure that estimates the input values using inputs–outputs of the system and IMU-INS measurement values is proposed in that study.

An observer/Kalman filter identification method for fault diagnosis in a small autonomous helicopter is proposed in [10]. The helicopter model under nominal conditions is found using input–output test values in an observer/Kalman filter. The differences between the designed model and the system outputs let the users determine the faults in the system. The model of the system, which is the helicopter model in this case, can be determined by Markov parameters and a Hankel matrix which has the same rank as the system.

In the given studies, different Kalman filter techniques (EKF, UKF, IMM-UKF, AMEKF), model-based methods such as using sensor signal characteristics, residual generators using polynomials, and NN, are used for state estimation and FDI. The methods vary due to the difficulty of the problem (multiple sensor fault detection, redundancy of sensors, etc.) where the FDI algorithm is used (aircraft, helicopter, etc.) and the seriousness of the fault. From these studies, we can understand that the Kalman filter technique can detect and isolate faults using different dynamic models and, in many studies, it is also used as a comparison test. In conclusion, this filtering technique is a useful method in determining faults in sensors.

In the literature survey, it is seen that many of the studies on the subject include fusing INS and IMU measurements with Global Positioning System (GPS) data. However, there are few studies on obtaining the best data for flight control from the air data system (ADS). In this study, the problem of obtaining the most appropriate data from ADS measurements is considered and ADS data FDI, determining the best data from the ADS, and estimation of the state subjects are investigated using high-accuracy GPS/INS measurements and the Kalman filtering technique as a basis.

7.2 Kalman Filter-Based Integrated ADS/GPS Navigation System

The indirect Kalman filtering technique is used in integrating two previously mentioned navigation systems' data. In this system, instead of estimating the system states, the errors between the system measurements are estimated [11].

In the integrated system scheme in Fig. 7.1, the inputs of the filter are speed measurement differences between the two different measurement systems. The speed errors of the ADS are estimated by the filter.

The system error vector with the required parameters is as follows:

$$x = \left[V_{eadsx}\ V_{eadsy}\ V_{eadsz} \right]^{\mathrm{T}} \tag{7.1}$$

Here, V_{eadsx}, V_{eadsy}, V_{eadsz} are the speed errors of the ADS, each in Cartesian coordinates. In the ADS, the true airspeed error is mainly the result of the wind

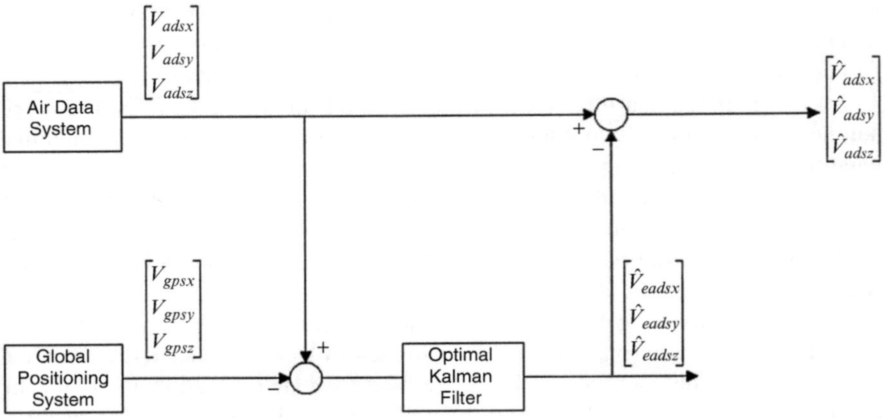

Fig. 7.1 Kalman filter-based air data system (ADS)/Global Positioning System (GPS) fusion scheme

speed. Therefore, choosing the ADS errors as the vector parameters shown in Eq. (7.1) is for the purpose of obtaining the true airspeed error, which is relatively high, and, thus, the wind speed.

The uncorrelated model, which is used in the exponential correlation function for the stagnant processes, will be used [11, 12]. This expression is in discrete and matrix form. This model is very suitable for simulation purposes [11]:

$$\begin{bmatrix} V_{eadsx}(k+1) \\ V_{eadsy}(k+1) \\ V_{eadsz}(k+1) \end{bmatrix} = \begin{bmatrix} 1-\beta_{V_{adsx}}T & 0 & 0 \\ 0 & 1-\beta_{V_{adsy}}T & 0 \\ 0 & 0 & 1-\beta_{V_{adsz}}T \end{bmatrix} \begin{bmatrix} V_{eadsx}(k) \\ V_{eadsy}(k) \\ V_{eadsz}(k) \end{bmatrix} + \begin{bmatrix} \left(\beta_{V_{adsx}}T - \frac{1}{2}\beta^2{}_{V_{adsx}}T^2\right) w_{V_{adsx}} \\ \left(\beta_{V_{adsy}}T - \frac{1}{2}\beta^2{}_{V_{adsy}}T^2\right) w_{V_{adsy}} \\ \left(\beta_{V_{adsz}}T - \frac{1}{2}\beta^2{}_{V_{adsz}}T^2\right) w_{V_{adsz}} \end{bmatrix} \quad (7.2)$$

where $\beta_{V_{adsx}}$, $\beta_{V_{adsy}}$, $\beta_{V_{adsz}}$ are the variables which are inverses of the true airspeed correlation time, $w_{V_{adsx}}$, $w_{V_{adsy}}$, $w_{V_{adsz}}$ are the Gauss distributed noises of the ADS speed measurements, and T is the sampling period.

Using the ADS and GPS speed measurement differences as the measurements observation vector in the Kalman Filter, the observation vector can be written as:

$$\begin{aligned} z_1(k) &= V_{eadsx} + v_{V_{adsx}} - v_{V_{gpsx}} \\ z_2(k) &= V_{eadsy} + v_{V_{adsy}} - v_{V_{gpsy}} \\ z_3(k) &= V_{eadsz} + v_{V_{adsz}} - v_{V_{gpsz}} \end{aligned} \quad (7.3)$$

In Eq. (7.3), V_{eadsx}, V_{eadsy}, V_{eadsz} are the true airspeed measurement errors of the ADS and, at the same time, are the wind speeds, and $\upsilon_{V_{adsx}}$, $\upsilon_{V_{adsy}}$, $\upsilon_{V_{adsz}}$ and $\upsilon_{V_{gpsx}}$, $\upsilon_{V_{gpsy}}$, $\upsilon_{V_{gpsz}}$ are the zero-mean Gauss noises of the ADS and GPS measurements, respectively. Namely, the difference between the GPS speed measurement and the ADS true airspeed measurement gives us the ADS speed error and the wind speed during flight. However, this wind speed information includes the random noises of both systems. The measurement statements of Eq. (7.3) can be written in matrix form as:

$$z(k) = \begin{bmatrix} z_1(k) \\ z_2(k) \\ z_3(k) \end{bmatrix} = \begin{bmatrix} V_{ADSX} - V_{GPSX} \\ V_{ADSY} - V_{GPSY} \\ V_{ADSZ} - V_{GPSZ} \end{bmatrix} = \underbrace{\begin{bmatrix} 1\,0\,0 \\ 0\,1\,0 \\ 0\,0\,1 \end{bmatrix}}_{H(k)} x(k) + \begin{bmatrix} \upsilon_{Vx} \\ \upsilon_{Vy} \\ \upsilon_{Vz} \end{bmatrix} \tag{7.4}$$

For integration of the ADS and INS/GPS data, the Kalman filtering equations below are used:

$$\begin{aligned} &\hat{x}(k/k) = \hat{x}(k/k-1) + K(k)\Delta(k) \\ &\Delta(k) = z(k) - H(k)\hat{x}(k/k-1) \\ &K(k) = P(k/k)\, H^T(k) R^{-1}(k) \\ &P(k/k) = (I - K(k)H(k))\, P(k/k-1) \\ &P(k/k-1) = \Phi(k,k-1)\, P(k-1/k-1)\, \Phi(k,k-1) \\ &\qquad\qquad + G(k,k-1)\, Q(k-1)\, G(k,k-1) \end{aligned} \tag{7.5}$$

where $\hat{x}(k/k)$ is the estimation value, I is the identity matrix, $K(k)$ is the gain value of the optimal filter, $P(k/k)$ is the estimation error correlation matrix, $P(k/k-1)$ is the extrapolation error correlation matrix, $H(k)$ is the measurement matrix, $\Delta(k)$ is the innovation process, $Q(k-1)$ is the system measurement correlation matrix, $G(k,k-1)$ is the noise transfer matrix, and $\Phi(k,k-1)$ is the system transfer matrix:

$$\Phi(k,k-1) = \begin{bmatrix} 1 - \beta_{V_{adsx}} T & 0 & 0 \\ 0 & 1 - \beta_{V_{adsy}} T & 0 \\ 0 & 0 & 1 - \beta_{V_{adsz}} T \end{bmatrix} \tag{7.6}$$

The initial conditions are:

$$\begin{aligned} &\hat{x}(0/0) = \overline{x(0)} \\ &P(0/0) = P(0) \end{aligned}$$

To start the Kalman filter process, the $x(0)$ and $P(0)$ initial conditions, system error correlation matrix $Q(k)$, and measurement error correlation matrix $R(k)$ must first be determined. To perform the filtering, we need system and measurement error models.

In Fig. 7.1, as we can see in the given integration scheme, the Kalman filter outputs are the ADS speed error estimation values, which are actually the wind speeds $\widehat{V}_{eadsx}$, $\widehat{V}_{eadsy}$, $\widehat{V}_{eadsz}$. These estimated values are subtracted from the ADS measurements to achieve the speeds $\widehat{V}_{adsx}$, $\widehat{V}_{adsy}$, $\widehat{V}_{adsz}$. We can use the following formulas for determining the speed values:

$$\begin{aligned}\widehat{V}_{adsx} &= V_{adsx} - \widehat{V}_{eadsx}\\ \widehat{V}_{adsy} &= V_{adsy} - \widehat{V}_{eadsy}\\ \widehat{V}_{adsz} &= V_{adsz} - \widehat{V}_{eadsz}\end{aligned} \tag{7.7}$$

7.3 Federated Kalman Filter-Based Integrated ADS and GPS/INS Data

7.3.1 Data Fusion Methodology

A standard central Kalman filter with central measurements and a covariance matrix can cause computational overload. In addition, any sensor fault can degrade the overall performance of the filter. To overcome these disadvantages, a noncentral Kalman filter is useful. A noncentral Kalman filter is a two-stage data-processing technique and is composed of one basic filter and one or more local filters [13–15]. Firstly, the local filters process the data optimally to obtain the best result using each set. Then, the central filter fuses local filter results and produces the most optimal value. The described method is an effective technique for integrating multiple sensor measurement sets. The federated Kalman filter has many advantages over the other noncentral filtering techniques [11, 13–15].

The federated Kalman filter-based navigation sensor fusion scheme, which fuses INS, GPS, and other navigation sensors, is given in Fig. 7.2.

The aforementioned federated Kalman filtering is a two-level data-processing technique. Local filters that process data from the sensors are fused in a federated filter. All the local filters estimate the same variables; thus, a correlation is present between them. A global optimal estimation and error variance can be described by the following equations:

$$\begin{aligned}P_m &= \left[P_1^{-1} + P_2^{-1} + \cdots + P_N^{-1}\right]^{-1}\\ X_{e_m} &= P_m\left[P_1^{-1}X_{e_1} + P_2^{-1}X_{e_2} + \cdots + P_N^{-1}X_{eN}\right]\end{aligned} \tag{7.8}$$

In the formulas, P_i is the covariance matrix of the ith filter and X_{e_i} is the estimation value of the ith filter.

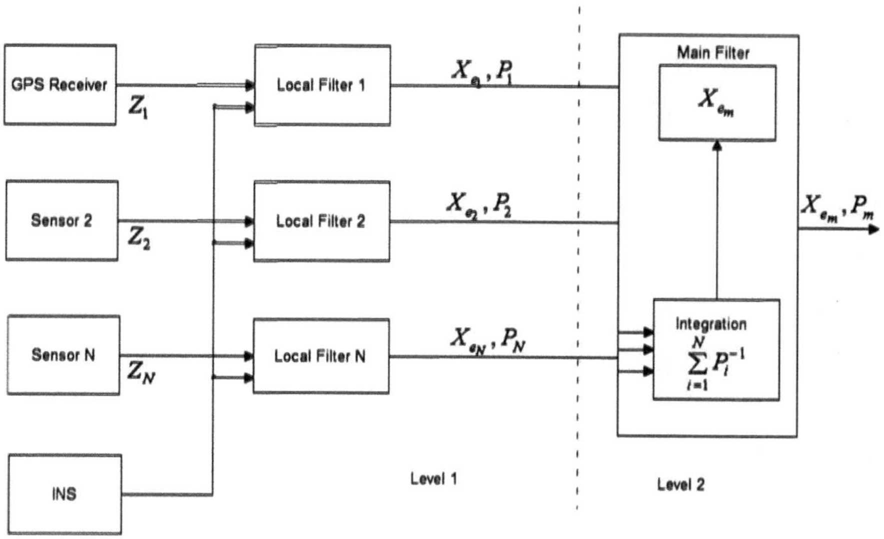

Fig. 7.2 Federated Kalman filter-based navigation sensor fusion scheme

7.3.2 *ADS and GPS/INS Data Fusion Based on FDI*

In this study, INS/GPS measurements, which have high accuracy, and ADS measurements, which have low accuracy but high frequency, are integrated using the Kalman filtering technique in order to obtain high-accuracy measurement data at high frequency. The designed system based on the indirect Kalman filtering technique is shown to be successful in finding the wind speed data using the known error or UAV dynamics of the system and determined statistical values.

In applications, generally, ADS and GPS or GPS/INS are integrated on the basis of a Kalman filter. A federated Kalman filter-based navigation sensor fusion scheme with FDI block to fuse the reliable data coming from the three ADS measurement sets and integrated GPS/INS (faulty data are determined and isolated) is shown in Fig. 7.3.

In the figure, it can be seen that the federated filter is composed of local filters that process ADS navigation measurements and GPS/INS measurements (position and velocity), together with the ADS data, FDI block for faulty data determination and isolation, and a main filter that fuses the estimates of the local filters after the elimination of faulty data. The main best variable estimates and error variance of the filters can be found using the equations given below [7, 9]:

$$\begin{aligned} P_m &= \left[P_1^{-1} + P_2^{-1} + P_3^{-1}\right]^{-1} \\ X_{e_m} &= P_m \left[P_1^{-1} X_{e_1} + P_2^{-1} X_{e_2} + P_3^{-1} X_{e_3}\right] \end{aligned} \tag{7.9}$$

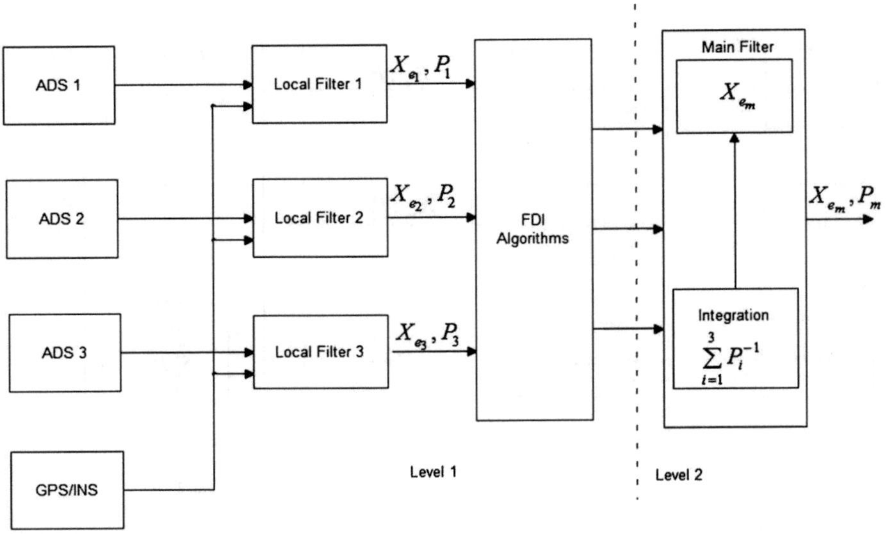

Fig. 7.3 Kalman filter-based ADS and GPS/inertial navigation system (INS) data fusion with fault detection and isolation (FDI)

The aforementioned algorithm can be used to derive the best data from the ADS measurements.

7.4 Sensor FDI Algorithms

The sensor FDI algorithms are presented in this section.

7.4.1 Statistical Test for Fault Detection

Two hypotheses are introduced:

H_0: the system operates normally
H_1: a fault occurs in the system

Using an innovation approach is suitable for detecting sensor faults [16, 17]. To detect failures changing the mean of the innovation sequence, the following statistical function can be used:

$$\beta(k) = \sum_{j=k-M+1}^{k} \tilde{\Delta}^{\mathrm{T}}(j)\tilde{\Delta}(j) \tag{7.10}$$

where $\tilde{\Delta}(j)$ is the normalized innovation sequence of the Kalman filter and M is the width of the sliding window.

The Kalman filter normalized innovation can be calculated as follows:

$$\tilde{\Delta}(k) = \left[H(k)P(k/k-1)H^{\mathrm{T}}(k) + R(k)\right]^{-1/2}\Delta(k) \tag{7.11}$$

This statistical function has a χ^2 distribution with Ms degrees of freedom, where s is the dimension of the state vector. If the level of significance, α, is selected as:

$$P\left\{\chi^2 > \chi^2_{\alpha,Ms}\right\} = \alpha; \quad 0 < \alpha < 1 \tag{7.12}$$

then the threshold value, $\chi^2_{\alpha,Ms}$, can be found. Hence, when the hypothesis H_1 is true, the statistical value of $\beta(k)$ will be greater than the threshold value $\chi^2_{\alpha,Ms}$, i.e.:

$$\begin{aligned} H_0 &: \beta(k) \le \chi^2_{\alpha,Ms}, \forall k \\ H_1 &: \beta(k) > \chi^2_{\alpha,Ms}, \exists k \end{aligned} \tag{7.13}$$

7.4.2 Fault Isolation Algorithm

If the fault is detected, then it is necessary to determine what sensor set is faulty. For this purpose, the s-dimensional sequence $\tilde{\Delta}$ is transformed into s one-dimensional sequences to isolate the faulty sensor, and for each one-dimensional sequence $\tilde{\Delta}_i\ (i = 1, 2, \ldots, s)$, the corresponding monitoring algorithm is run. The statistics of the faulty sensor set are assumed to be affected much more than those of the other sensors. Let the statistics be denoted as $\xi_i(k)$. When $\max\{\xi_i(k)/i = 1, 2, \ldots, s\} = \xi_m(k)$ for $i \ne j$ and $\xi_i(k) \ne \xi_j(k)$, it is judged that the sensor set has failed.

Let the statistics that is a rate of sample and theoretical variances, $\widehat{\sigma}_i^2 \big/ \sigma_i^2$, be used to verify the variances of one-dimensional innovation sequences $\tilde{\Delta}_i(k), i = 1, 2, \ldots, s$. When $\tilde{\Delta}_i \sim N(0, \sigma_i)$, it is known that [18]:

$$\frac{\nu_i}{\sigma_i^2} \sim \chi^2_{M-1}, \forall i, i = 1, 2, \ldots, s \tag{7.14}$$

where $\nu_i = (M-1)\widehat{\sigma}_i^2$.

As $\sigma_i^2 = 1$ for the normalized innovation sequence, it follows that:

$$\nu_i \sim \chi^2_{M-1}, \forall i, i = 1, 2, \ldots, s. \tag{7.15}$$

Using Eq. (7.15), it can be proved that any change in the mean of the normalized innovation sequence can be detected [19]. When a fault affecting the mean or variance of the innovation sequence occurs in the system, the statistics ν_i exceeds the threshold value $\chi^2_{\alpha,M-1}$, depending on the level of significance α and degrees of freedom $(M - 1)$.

7.5 Simulation Results for Indirect Kalman Filter-Based ADS and GPS/INS Data Fusion

Three ADS sets are taken into consideration. The sets are composed of values taken from an AOA sensor, sideslip angle sensor (AOS), and speed values determined from true airspeed sensor (TAS) measurements. There are three AOA, two AOS, and three TAS sensors. The first set is built using AOA sensor 1 (AOA1), AOS sensor 1 (AOS1), and TAS sensor 1 (TAS1). The second set includes AOA sensor 2 (AOA2), AOS sensor 2 (AOS2), and TAS sensor 2 (TAS2). The third set is composed of AOA sensor 3 (AOA3), AOS2 (as there are two different AOS sensors), and TAS sensor 3 (TAS3).

Three different cases are investigated in the simulations [20]:

1. TAS1 bias = 10 knots (+10 knots disturbance on the first set's true airspeed data after 80 s)
2. AOA3 bias = 5° (constant +5° bias on the third set's angle of attack data after 80 s)
3. AOS1 bias =5° (constant +5° bias on the first set's sideslip angle data after 80 s)

The faulty data are eliminated using the statistical test as described earlier. After that, data from another set that are known to be true are used in place of the faulty data. For example, when the TAS1 data is determined as faulty, TAS2 will be used instead to make set 1 working. Of course, as there are only two sideslip angle sensors, when one AOS value is wrong, there is only one other value to replace it. In the second simulation, the AOA3 value is faulty and it is replaced by the AOA1 value as soon as the system detects the fault.

These sets are simulated with systems that isolate the faults and eliminate the faulty data (Figs. 7.4, 7.5, 7.6, and 7.7) and the systems that do not perform isolation (Figs. 7.8, 7.9, 7.10, and 7.11). In the figures, "beta" denotes the sideslip angle and "KTAS" refers to the knots true airspeed.

7.5.1 Results with Fault Isolation

The case for the system with FDI when the TAS1 bias equal to 10 knots occurs at the 80th second is investigated. Using the statistical tests, faulty data can be determined and isolated. The variables AOA and TAS estimated using the filters are given in Figs. 7.4 and 7.5, respectively.

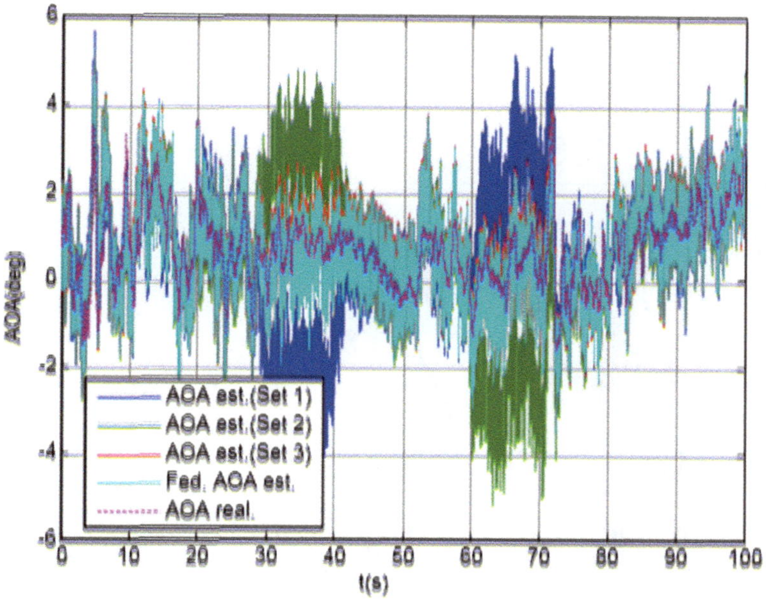

Fig. 7.4 Angle of attack (AOA) estimation values with fault isolation [when true airspeed sensor 1 (TAS1) bias = 10 knots]

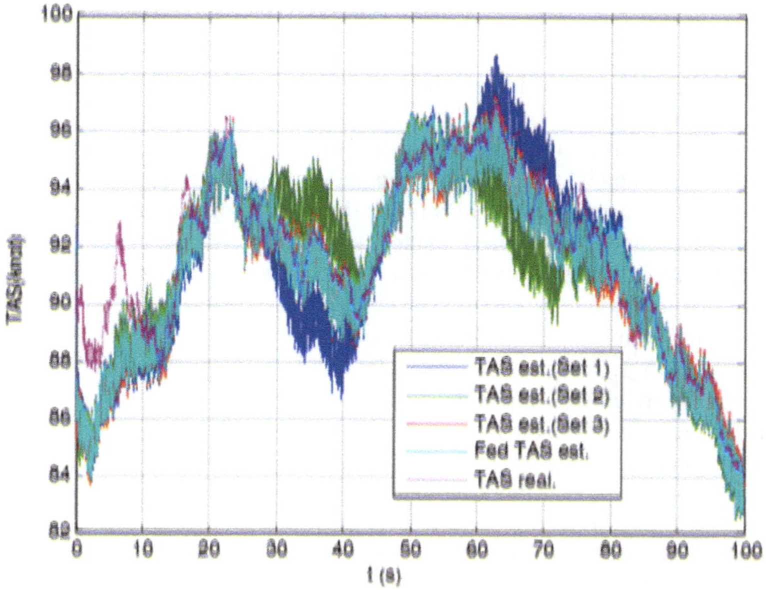

Fig. 7.5 TAS estimation values with fault isolation (when TAS1 bias = 10 knots)

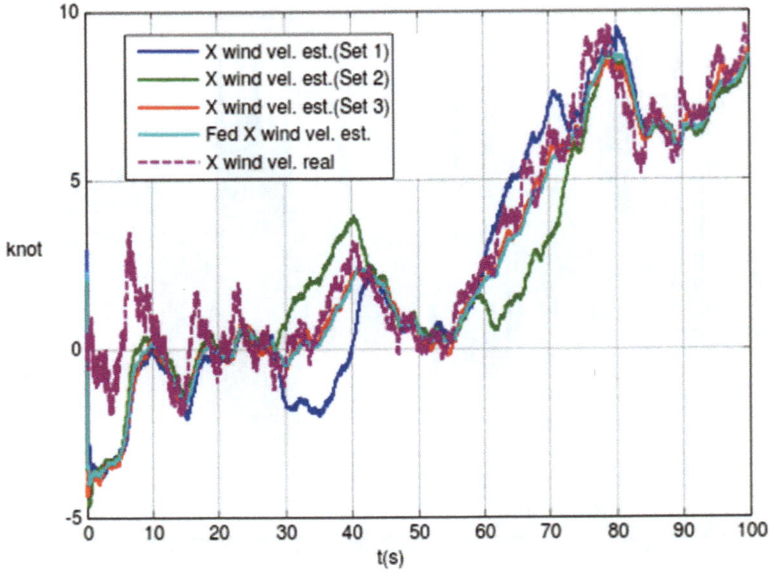

Fig. 7.6 Wind velocity estimation values in the *x*-direction with fault isolation (when TAS1 bias = 10 knots)

In Fig. 7.6, the ADS speed error estimation in the *x*-direction (wind velocity estimation in the *x*-direction) and real wind velocity in the *x*-direction are shown together.

It can be seen that the wind velocity in the *x*-direction is accurately estimated by the federated Kalman filter, even though the first and second sets do not give good results at different time intervals. We can also tell that some of the measurement values are eliminated by the system that includes the federated filter technique.

Secondly, the case where the bias in the third set's angle of attack value is $+5°$ after the 80th second is considered. In these simulations, it must be remembered that the isolation algorithm is also working and that the faulty data are eliminated.

Again, for this set, fault isolation is carried out and AOA and TAS estimation values are shown in Figs. 7.7 and 7.8, respectively.

As seen in Figs. 7.7 and 7.8, the estimated values of AOA and TAS via the federated filter are close to the real values.

Thirdly, the case where the bias on the first set's sideslip angle value is $+5°$ after the 80th second is considered. In these simulations, it must be remembered that the isolation algorithm is also working and that the faulty data are eliminated. Wind speed values in the *y*-direction for this case are presented in Fig. 7.9.

As seen from Fig. 7.9, the wind velocity in the *y*-direction is accurately estimated by the federated Kalman filter. Similar simulation results are obtained for the rest of the parameters of the UAV ADS.

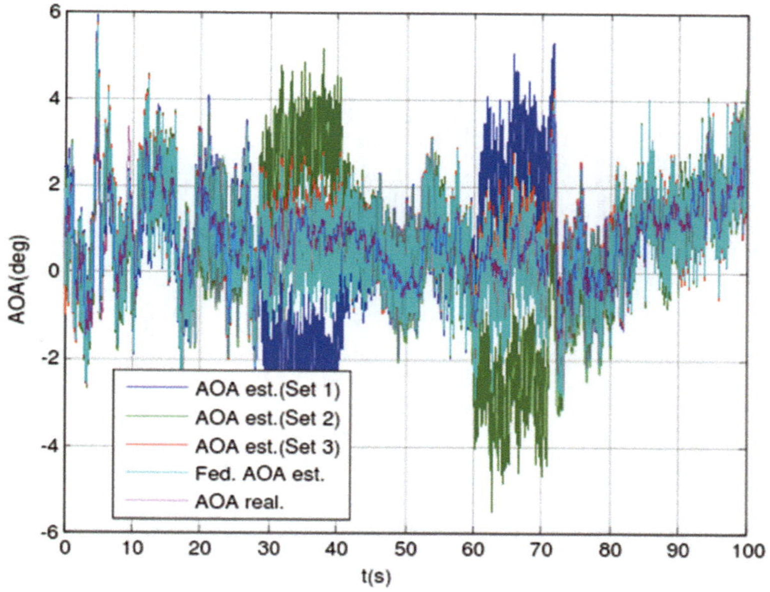

Fig. 7.7 AOA estimation values with fault isolation (when AOA3 bias = +5°)

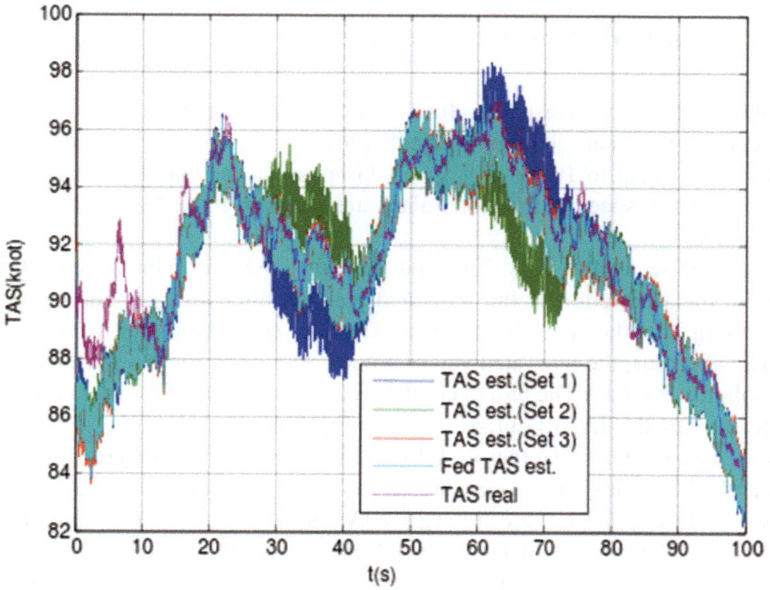

Fig. 7.8 TAS estimation values with fault isolation (when AOA3 bias = +5°)

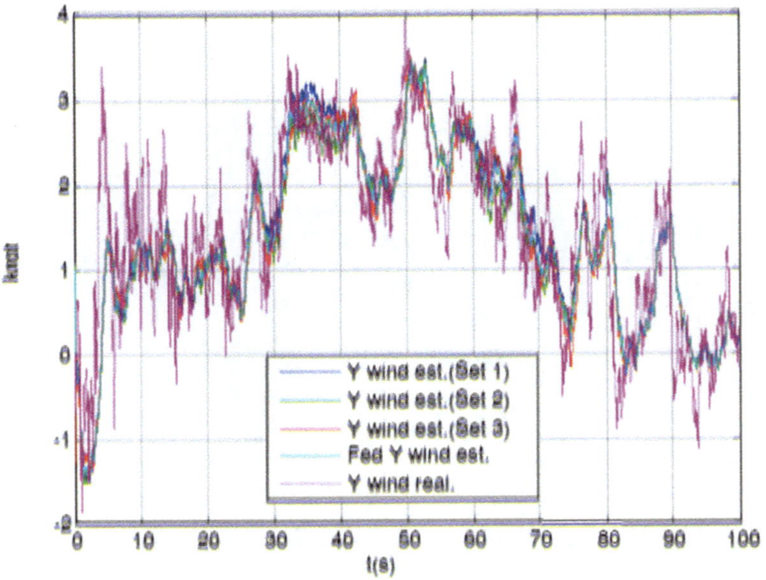

Fig. 7.9 Wind speed values in the *y*-direction with fault isolation (when AOS1 bias = +5°)

7.5.2 Results Without Fault Isolation

The federated filter is investigated for the same three cases. If isolation is not made, the federated filter can give faulty results [20]. Again, firstly, the results when the TAS1 bias is equal to 10 knots is given. Graphs that include the AOA, TAS, and *x*-direction wind speed estimation values are given in Figs. 7.10, 7.11, and 7.12, respectively.

As can be seen from Figs. 7.11 and 7.12, the TAS and wind speed estimation values in the *x*-direction of the federated Kalman filter diverge. The cause is the data coming from the first estimation set. The filter takes the values from the first set into account; thus, the federated values are not very good for this reason (no faulty data isolation is made).The federated filter estimated values of AOA are not affected significantly by the TAS1 fault, as shown in Fig. 7.10.

Next, the case where the bias in AOA in set 3 is +5° is considered. The estimated AOA and TAS values for this case are given in Figs. 7.13 and 7.14, respectively.

As seen in Fig. 7.13, the estimated AOA values of the federated Kalman filter also diverge. The values with errors, which are not isolated and used in calculations, cause this divergence. The federated filter estimated values of TAS are not affected significantly by the AOA3 fault, as shown in Fig. 7.14.

Wind speed estimation results in the *y*-direction when the AOS1 bias = 5° and fault isolation is absent are shown in Fig. 7.15.

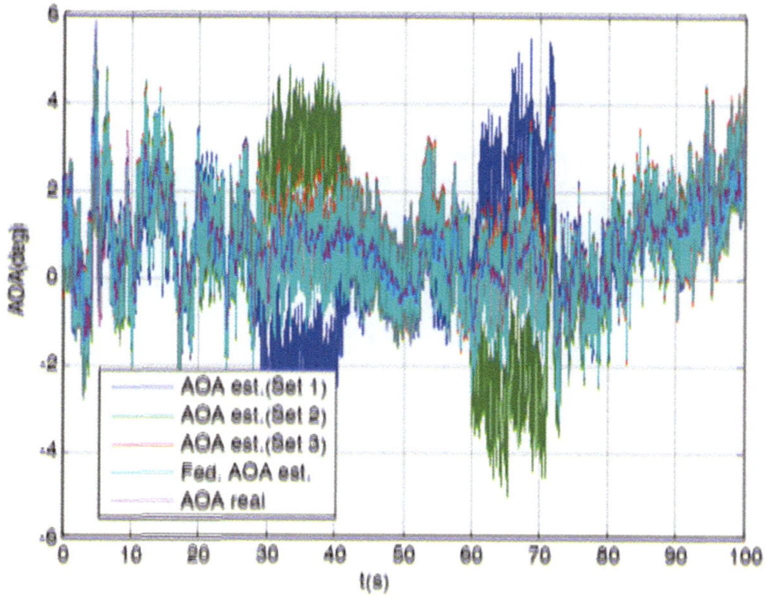

Fig. 7.10 AOA estimation values without fault isolation (when TAS1 bias = 10)

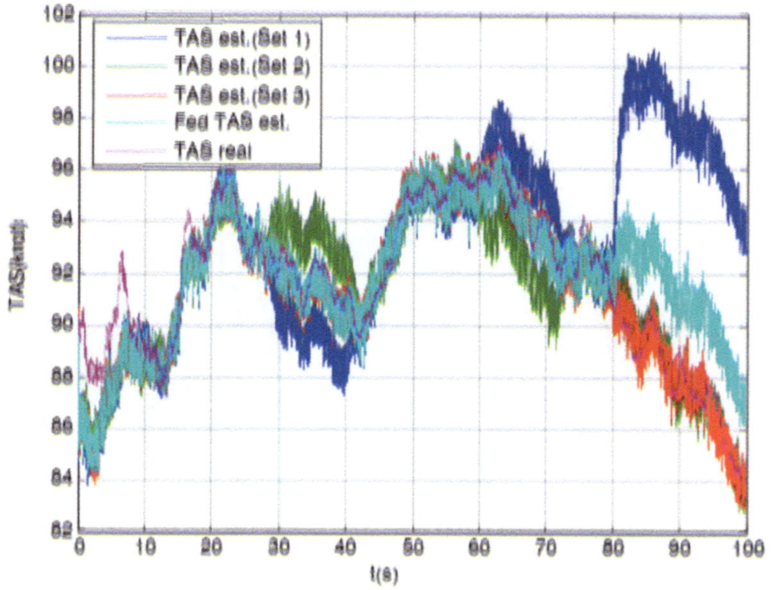

Fig. 7.11 TAS estimation values without fault isolation (when TAS1 bias = 10 knots)

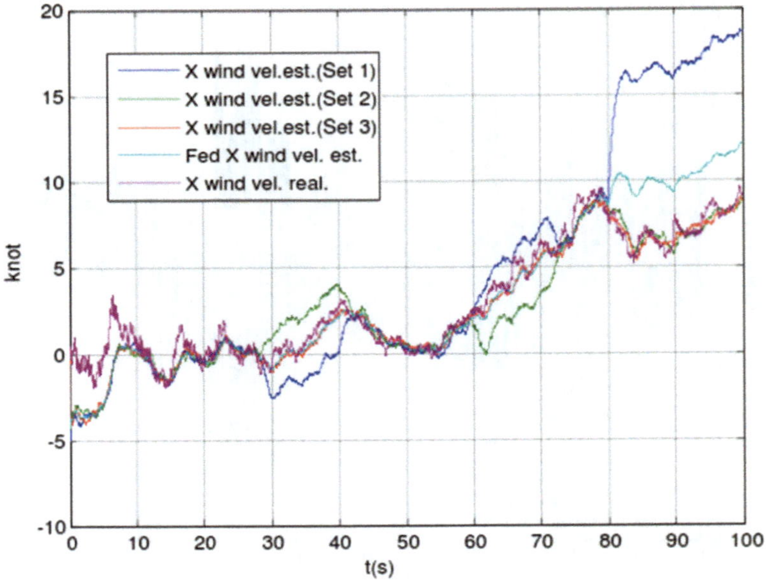

Fig. 7.12 Wind velocity estimation values in the *x*-direction without fault isolation (when TAS1 bias = 10)

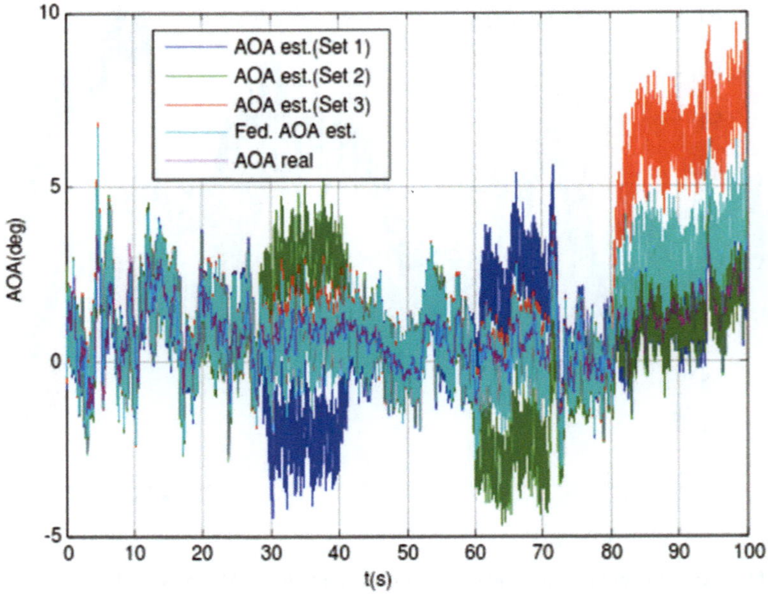

Fig. 7.13 AOA estimation values without fault isolation (AOA3 bias = +5°)

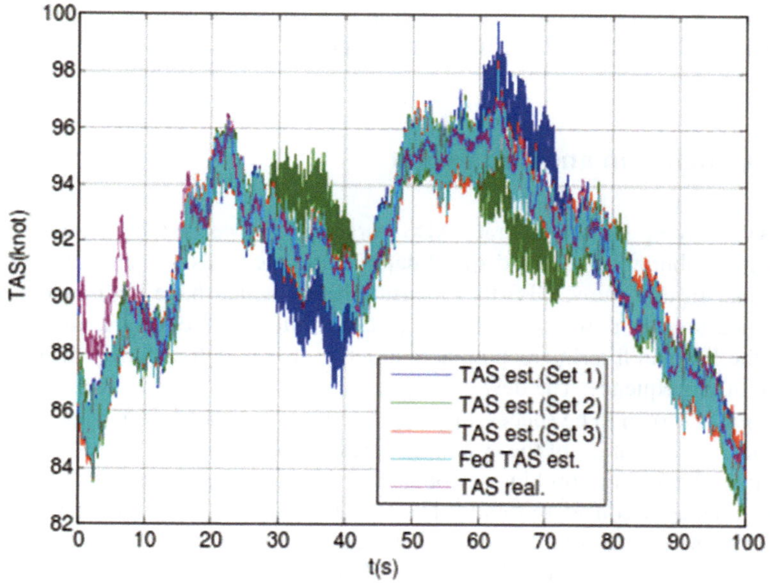

Fig. 7.14 TAS estimation values without fault isolation (AOA3 bias = +5°)

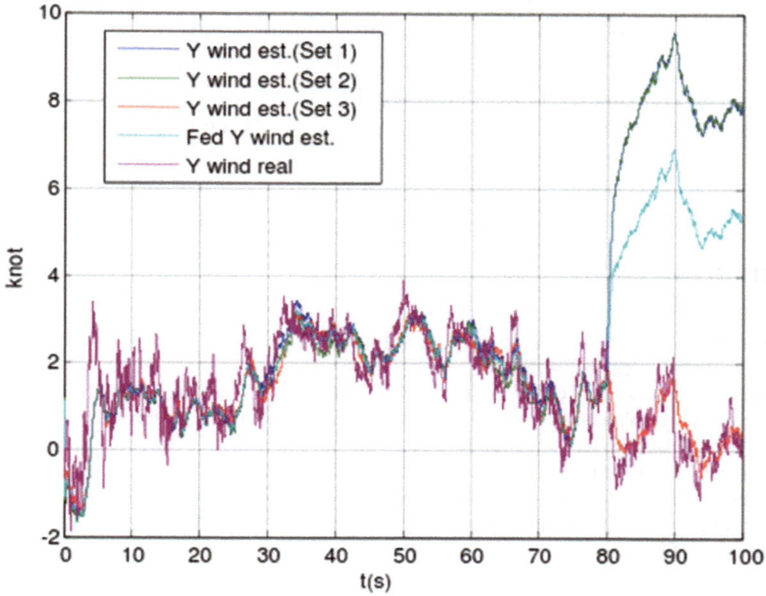

Fig. 7.15 Wind speed estimation results in the y-direction without fault isolation (when AOS1 bias = 5°)

As can be seen from the presented graphs, wind speed estimation values in the y-direction of the federated Kalman filter diverge.

7.6 Conclusion and Discussion

In UAVs, having highly accurate ADS parameters is needed for the purposes of control. To obtain the required speed data, fusing the measurements of the system is a must. In this study, GPS/INS measurements, which have high accuracy, and ADS measurements, which have low accuracy but high frequency, are integrated using the Kalman filtering technique in order to obtain high-accuracy measurement data at high frequency. The designed system based on the indirect Kalman filtering technique is shown to be successful in finding the wind speed data using the known error or UAV dynamics of the system and determined statistical values.

In this chapter, the problem of obtaining the most appropriate data from ADS measurements is considered and ADS data FDI, determining the best data from the ADS, and estimation of the state subjects are investigated using high-accuracy GPS/INS measurements and the Kalman filtering technique as a basis.

FDI algorithms are developed and diagnostic tests performed for faulty cases. The mentioned tests are based on the statistical properties of the normalized innovation process of the Kalman filter.

To fuse the data originating from the different measurement sets, a federated Kalman filter is presented. The mentioned algorithms can be used to derive the best data from the ADS measurements. The best data are produced using the FDI algorithm and leaving the faulty data out before fusing with the federated filter. As a result, the air data parameters are estimated by the federated Kalman filter very well and the estimated values converge to the real values.

References

1. Jayaram S (2010) A new fast converging Kalman filter for sensor fault detection and isolation. Sensor Rev 30(3):219–224
2. Bryson M, Sukkarieh S (2004) Vehicle model aided inertial navigation for a UAV using low-cost sensors. In: Proceedings of the 2004 Australasian conference on robotics and automation, Canberra, Australia, December 2004
3. Mosallae M, Salahshoor K (2008) Sensor fault detection using adaptive modified extended Kalman filter based on data fusion technique. In: Proceedings of the 4th international conference on information and automation for sustainability (ICIAFS08), Colombo, Sri Lanka, December 2008
4. Samara PA, Sakellariou JS, Fouskitakis GN, Fassois SD (2003) Detection of sensor abrupt faults in aircraft control systems. In: Proceedings of the 2003 IEEE conference on control applications (CCA 2003), Istanbul, Turkey, June 2003
5. Bonfe M, Castaldi P, Geri W, Simani S (2006) Fault detection and isolation for on-board sensors of a general aviation aircraft. Int J Adapt Control Signal Process 20:381–408

6. Samy I, Postlethwaite I, Gu DW (2009) Sensor fault detection and accommodation using neural networks with application to a non-linear unmanned air vehicle model. In: Proceedings of the IMechE Part G: J Aerosp Eng, vol 224
7. Samara PA, Fouskitakis GN, Sakellariou JS, Fassois D (2008) A statistical method for the detection of sensor abrupt faults in aircraft control systems. IEEE Trans Control Syst Technol 16(4):789–798
8. Cork L, Walker R (2007) Sensor fault detection for UAVs using a nonlinear dynamic model and the IMM-UKF algorithm. IEEE Inf Decis Control, pp 230–235
9. Marzat J, Piet-Lahanier H, Damongeot F, Walter E (2010) Fault diagnosis for nonlinear aircraft based on control-induced redundancy. In: Proceedings of the conference on control and fault-tolerant systems (SysTol), Nice, France, October 2010
10. Heredia G, Ollero A (2009) Sensor fault detection in small autonomous helicopters using observer/Kalman filter identification. In: Proceedings of the 2009 IEEE international conference on mechatronics, Malaga, Spain, April 2009
11. Hajiyev C (2010) Experimental data processing methods and engineering applications. Nobel Publication and Distribution Inc., Ankara (in Turkish)
12. Grishin YP, Kazarinov YM (1985) Fault-tolerant dynamic systems. Radio i Svyaz, Moscow (in Russian)
13. Carlson NA (1990) Federated square root filter for decentralized parallel processors. IEEE Trans Aerosp Electron Syst 26:517–525
14. Carlson NA, Berarducci MP (1994) Federated Kalman filter simulation results. Navigation 41:297–321
15. Hajiyev C, Tutucu A (2001) INS/GPS integration using parallel Kalman filtering. In: Proceedings of the 15th IFAC symposium on automatic control in aerospace, Bologna/Forli, Italy, pp 453–458
16. Willsky AS (1976) A survey of design methods for failure detection in dynamic systems. Automatica 12(6):601–611
17. Hajiyev C, Caliskan F (2003) Fault diagnosis and reconfiguration in flight control systems. Kluwer Academic Publishers, Boston
18. Hajiyev C (2006) Innovation approach based measurement error self-correction in dynamic systems. Measurement 39:585–593
19. Hajiyev C, Caliskan F (2005) Sensor and control surface/actuator failure detection and isolation applied to F-16 flight dynamic. Aircr Eng Aerosp Technol 77(2):152–160
20. Hajiyev Ch, Vural SY, Kargin V (2012) Fault detection, isolation and federated filtration in UAV air data system. In: Proceedings of the national aerospace conference, 12–14 September 2012. Turkish Air Force Academy, Istanbul, UHUK-2012-043, 16 (in Turkish)

Chapter 8
Stability Analysis for Unmanned Aerial Vehicles

8.1 Trimming

The equations of motion for unmanned aerial vehicles (UAVs) are given in Chap. 2. In this chapter, the equations of motion are further examined and their linearization is discussed in detail for the stability analysis. In this procedure, the first step is to investigate the trim conditions.

8.1.1 Trim Point

A nonlinear system can be described as:

$$\dot{x} = f(x, u), \tag{8.1}$$

where u represents the system control inputs and x represents the states of the system. The trim condition for the system can be found as:

$$f(\widehat{x}, \widehat{u}) = 0. \tag{8.2}$$

If the system is in equilibrium with the given states and control inputs, the variables do not change. Stable (trimmed) flight condition for a UAV is an example of such a situation. In this case, we can say that a subspace of the UAV states is in equilibrium. In the literature, this case is called the trim condition. Wings-level, fixed-altitude turn, and climb–descent maneuvers are different examples of trim conditions.

C. Hajiyev et al., *State Estimation and Control for Low-cost Unmanned Aerial Vehicles*, DOI 10.1007/978-3-319-16417-5_8

8.1.2 Linearization Around a Steady-State Condition

The states of a UAV can be calculated for different trim conditions using a minimization function. The linearization of the equations is necessary for controller design.

We can find the linearized equations of motion for a UAV starting with the equations given in Chap. 2.

Around the equilibrium point, we have:

$$\dot{\widehat{x}} = f\left(\widehat{x}, \widehat{u}\right) \tag{8.3}$$

A change around the equilibrium point can be represented as:

$$\tilde{x} = x - \widehat{x}, \tag{8.4}$$

$$\dot{\tilde{x}} = \dot{x} - \dot{\widehat{x}}, \tag{8.5}$$

$$\dot{\tilde{x}} = f\left(x, u\right) - f\left(\widehat{x}, \widehat{u}\right), \tag{8.6}$$

$$\dot{\tilde{x}} = f\left(x + \tilde{x} - \tilde{x}, u + \tilde{u} - \tilde{u}\right) - f\left(\widehat{x}, \widehat{u}\right), \tag{8.7}$$

$$\dot{\tilde{x}} = f\left(\widehat{x} + \tilde{x}, \widehat{u} + \tilde{u}\right) - f\left(\widehat{x}, \widehat{u}\right). \tag{8.8}$$

With Taylor series approximation, we can find:

$$\dot{\tilde{x}} = f\left(\widehat{x}, \widehat{u}\right) + \frac{\partial f\left(\widehat{x}, \widehat{u}\right)}{\partial x}\tilde{x} + \frac{\partial f\left(\widehat{x}, \widehat{u}\right)}{\partial u}\tilde{u} + \cdots - f\left(\widehat{x}, \widehat{u}\right), \tag{8.9}$$

Then, the changes in states can be given with the following equation:

$$\dot{\tilde{x}} \cong \frac{\partial f\left(\widehat{x}, \widehat{u}\right)}{\partial x}\tilde{x} + \frac{\partial f\left(\widehat{x}, \widehat{u}\right)}{\partial u}\tilde{u}. \tag{8.10}$$

In order to linearize the equations, we need to solve for $\frac{\partial f}{\partial x}$ and $\frac{\partial f}{\partial u}$ at the equilibrium point.

8.2 Derivation of the Transfer Functions

8.2.1 State Equations

The longitudinal and lateral motions can be handled separately. The variables that we choose to describe the longitudinal and lateral motions of the UAV are h, V, α, u, w, q, θ and β, v, p, r, Φ, ψ, respectively.

The longitudinal equations of motion for the UAV can be found by taking the lateral variables as zero and using definitions of $\alpha = \arctan\left(\frac{w}{u}\right)$ and $V_t = \sqrt{u^2 + w^2}$, where V_t is the true airspeed.

The differentiation matrix for the states can be found as:

$$\frac{\partial f}{\partial x} = \begin{bmatrix} \left(\frac{\partial \dot{u}}{\partial u}\right), \left(\frac{\partial \dot{u}}{\partial w}\right), \left(\frac{\partial \dot{u}}{\partial q}\right), \left(\frac{\partial \dot{u}}{\partial \theta}\right), \left(\frac{\partial \dot{u}}{\partial h}\right) \\ \left(\frac{\partial \dot{w}}{\partial u}\right), \left(\frac{\partial \dot{w}}{\partial w}\right), \left(\frac{\partial \dot{w}}{\partial q}\right), \left(\frac{\partial \dot{w}}{\partial \theta}\right), \left(\frac{\partial \dot{w}}{\partial h}\right) \\ \left(\frac{\partial \dot{q}}{\partial u}\right), \left(\frac{\partial \dot{q}}{\partial w}\right), \left(\frac{\partial \dot{q}}{\partial q}\right), \left(\frac{\partial \dot{q}}{\partial \theta}\right), \left(\frac{\partial \dot{q}}{\partial h}\right) \\ \left(\frac{\partial \dot{\theta}}{\partial u}\right), \left(\frac{\partial \dot{\theta}}{\partial w}\right), \left(\frac{\partial \dot{\theta}}{\partial q}\right), \left(\frac{\partial \dot{\theta}}{\partial \theta}\right), \left(\frac{\partial \dot{\theta}}{\partial h}\right) \\ \left(\frac{\partial \dot{h}}{\partial u}\right), \left(\frac{\partial \dot{h}}{\partial w}\right), \left(\frac{\partial \dot{h}}{\partial q}\right), \left(\frac{\partial \dot{h}}{\partial \theta}\right), \left(\frac{\partial \dot{h}}{\partial h}\right) \end{bmatrix} \tag{8.11}$$

Similarly, the differentiation matrix for the control inputs can be derived and the longitudinal state and control matrices—A_{lon} and B_{lon}, respectively—can be found. The equations for the longitudinal stability coefficients are given below:

$$X_u = \frac{u\rho S}{m}\left[c_{xo} + c_{x\alpha}\alpha + c_{x\delta_e}\delta e\right] - \frac{\rho S w c_{x\alpha} u}{2m} + \frac{\rho S\bar{c} c_{xq} u q}{2mV} - \frac{\rho S_{\text{prop}} c_{\text{prop}} u}{m} \tag{8.12}$$

$$X_w = -q + \frac{u\rho S}{m}\left[c_{xo} + c_{x\alpha}\alpha + c_{x\delta_e}\delta e\right] - \frac{\rho S V c_{x\alpha} u}{2m} + \frac{\rho S\bar{c} c_{xq} w q}{2mV} - \frac{\rho S_{\text{prop}} c_{\text{prop}} w}{m} \tag{8.13}$$

$$X_q = -w + \frac{\rho S V c_{xq} q \bar{c}}{2m} \tag{8.14}$$

$$X_{\delta e} = \frac{\rho V^2 c_{x\delta e} S}{2m} \tag{8.15}$$

$$X_{\delta t} = \frac{\rho S_{\text{prop}} c_{xp} \delta_a h}{m} \tag{8.16}$$

$$Z_u = q + \frac{u\rho S}{m}\left[c_{zo} + c_{z\alpha}\alpha + c_{z\delta_e}\delta e\right] - \frac{\rho S c_{z\alpha} w}{2m} + \frac{\rho S\bar{c} c_{zq} u q}{mV} \tag{8.17}$$

$$Z_w = \frac{w\rho S}{m}\left[c_{zo} + c_{z\alpha}\alpha + c_{z\delta_e}\delta e\right] - \frac{\rho S c_{z\alpha} u}{2m} + \frac{\rho S\bar{c} c_{zq} u q}{mV} \tag{8.18}$$

$$Z_q = u + \frac{\rho S c_{zq} V\bar{c}}{2m} \tag{8.19}$$

$$Z_q = \frac{\rho S c_{z\delta e} V^2}{2m} \tag{8.20}$$

$$M_u = \frac{u\rho S\overline{c}}{J_y}\left[c_{mo} + c_{m\alpha}\alpha + c_{m\delta e}\delta e\right] - \frac{\rho S w\overline{c} c_{m\alpha} u}{2J_y} + \frac{\rho S\overline{c}^2 c_{mq} uq}{2J_y V} \tag{8.21}$$

$$M_w = \frac{w\rho S\overline{c}}{J_y}\left[c_{mo} + c_{m\alpha}\alpha + c_{m\delta e}\delta e\right] + \frac{\rho S\overline{c} c_{m\alpha} w}{2J_y} + \frac{\rho S\overline{c}^2 c_{mq} wq}{2J_y V} \tag{8.22}$$

$$M_q = \frac{\rho S c_{mqe} V\overline{c}}{2J_y} \tag{8.23}$$

$$M_{\delta e} = \frac{\rho S c_{m\delta e} V^2 \overline{c}}{2J_y} \tag{8.24}$$

The derivation procedure for the lateral equations of motions and, therefore, the lateral state and control matrices—A_{lat} and B_{lat}, respectively—are similar. The equations for the lateral stability coefficients are given below:

$$Y_v = \frac{v\rho bS}{4mV_{\text{air}}}\left[c_{yp}p + c_{yr}r\right] + \frac{\rho S v}{m}\left(c_{yo} + c_{y\beta}\beta + c_{y\delta a}\delta a + c_{y\delta r}r\right) + \frac{\rho S c_{y\beta}}{2m}\sqrt{u^2 + w^2} \tag{8.25}$$

$$Y_p = w + \frac{\rho S V_{air} b}{4m} c_{yp} \tag{8.26}$$

$$Y_r = -u + \frac{\rho S V_{air} b}{4m} c_{yr} \tag{8.27}$$

$$Y_{\delta q} = \frac{\rho S {V_{air}}^2}{2m} c_{y\delta a} \tag{8.28}$$

$$Y_{\delta r} = \frac{\rho S {V_{air}}^2}{2m} c_{y\delta r} \tag{8.29}$$

$$L_v = \frac{v\rho b^2 S}{8mV_{\text{air}}}\left[c_{pp}p + c_{pr}r\right] + \frac{\rho S b v}{2m}\left(c_{po} + c_{p\beta}\beta + c_{p\delta a}\delta a + c_{p\delta r}\delta r\right) + \frac{\rho S b c_{p\beta}}{4m}\sqrt{u^2 + w^2} \tag{8.30}$$

$$L_p = \Gamma_1 q + \frac{\rho S {V^2}_{\text{air}} b}{8m} c_{pp} \tag{8.31}$$

$$L_r = -\Gamma_2 q + \frac{\rho S {V^2}_{\text{air}} b}{8m} c_{pr} \tag{8.32}$$

$$L_{\delta q} = \frac{\rho S b V_{\text{air}}^{2}}{4m} c_{p\delta a} \tag{8.33}$$

$$L_{\delta r} = \frac{\rho S b V_{\text{air}}^{2}}{4m} c_{p\delta r} \tag{8.34}$$

$$\begin{aligned} N_v &= \frac{v\rho b^2 S}{8mV_{\text{air}}}\left[c_{rp}p + c_{rr}r\right] + \frac{\rho S b v}{2m}\left(c_{ro} + c_{r\beta}\beta + c_{r\delta a}\delta a + c_{r\delta r}\delta r\right) \\ &+ \frac{\rho S b c_{r\beta}}{4m}\sqrt{u^2 + w^2} \end{aligned} \tag{8.35}$$

$$N_p = \Gamma_3 q + \frac{\rho S V^2{}_{\text{air}} b}{8m} c_{rp} \tag{8.36}$$

$$N_r = -\Gamma_4 q + \frac{\rho S V^2{}_{\text{air}} b}{8m} c_{rr} \tag{8.37}$$

$$N_{\delta q} = \frac{\rho S b V_{\text{air}}^{2}}{4m} c_{r\delta a} \tag{8.38}$$

$$N_{\delta r} = \frac{\rho S b V_{\text{air}}^{2}}{4m} c_{r\delta r} \tag{8.39}$$

The next step is to give the linearized state equations for the UAV. The equations in the state space form are:

$$\begin{pmatrix} \dot{u} \\ \dot{w} \\ \dot{q} \\ \dot{\theta} \\ \dot{h} \end{pmatrix} = \begin{bmatrix} (Xu) & (Xw) & (Xq) & (-g\cos(\theta)) & 0 \\ (Zu) & (Zw) & (Zq) & (-g\sin(\theta)) & 0 \\ Mu & Mw & Mq & 0 & 0 \\ 0 & 0 & 1 & 0 & 0 \\ -\sin(\theta) & -\cos(\theta) & 0 & t & 0 \end{bmatrix} \begin{pmatrix} u \\ w \\ q \\ \theta \\ h \end{pmatrix} + \begin{bmatrix} X_{\delta e} & X_{\delta t} \\ Z_{\delta e} & 0 \\ M_{\delta e} & 0 \\ 0 & 0 \\ 0 & 0 \end{bmatrix} \begin{bmatrix} \delta_e \\ \delta_t \end{bmatrix} \tag{8.40}$$

$$t = u\cos(\theta) + w\sin(\theta) \tag{8.41}$$

$$\begin{pmatrix} \dot{v} \\ \dot{p} \\ \dot{r} \\ \dot{\varphi} \\ \dot{\psi} \end{pmatrix} = \begin{bmatrix} (Yv) & v & -u & g\cos(\theta) & 0 \\ (Lv) & (Lp) & (Lr) & 0 & 0 \\ (Nv) & (Np) & (Nr) & 0 & 0 \\ 0 & 1 & \tan(\theta) & 0 & 0 \\ 0 & 0 & \sec(\theta) & 0 & 0 \end{bmatrix} \begin{pmatrix} v \\ p \\ r \\ \varphi \\ \psi \end{pmatrix} + \begin{bmatrix} Y_{\delta a} & Y_{\delta r} \\ L_{\delta a} & L_{\delta r} \\ N_{\delta a} & N_{\delta r} \\ 0 & 0 \\ 0 & 0 \end{bmatrix} \begin{bmatrix} \delta_a \\ \delta_r \end{bmatrix} \tag{8.42}$$

Here, ten states are used to describe the lateral and longitudinal motions of the UAV. The heading angle equation is also added to the system.

Finally, the linearized equations of motion for the Zagi UAV are found as below. These equations will be used when we design the various controller algorithms that are presented in the next chapters of this book. The values given in the matrices A_{lon}, A_{lat}, B_{lon}, and B_{lat} are for the trim conditions. The longitudinal equations are:

$$\dot{x} = Ax + Bu \tag{8.43}$$

$$A_{\text{lon}} = \begin{bmatrix} -0.3356 & 1.3181 & -1.9276 & -9.6610 & 0 \\ -1.7916 & -3.9003 & 9.8215 & -1.7035 & 0 \\ 0.7020 & -3.5375 & -11.3920 & 0 & 0 \\ 0 & 0 & 1.0000 & 0 & 0 \\ -0.1736 & -0.9848 & 0 & 17.4865 & 0 \end{bmatrix} \tag{8.44}$$

$$B_{\text{lon}} = \begin{bmatrix} -0.7436 & 6.8728 \\ 3.7855 & 0 \\ 47.9170 & 0 \\ 0 & 0 \\ 0 & 0 \end{bmatrix} \tag{8.45}$$

$$u_{\text{lon}} = \begin{bmatrix} \delta e \\ \delta t \end{bmatrix} \tag{8.46}$$

$$X_{\text{lon}} = \begin{bmatrix} u \; w \; q \; \theta \; h \end{bmatrix}^{\mathrm{T}} \tag{8.47}$$

On the other hand, the lateral equations are:

$$A_{\text{lat}} = \begin{bmatrix} -1.0502 & 1.9276 & -9.8215 & 9.6610 & 0 \\ -1.2213 & -1.9155 & 1.0096 & 0 & 0 \\ 1.7255 & 0.0919 & -1.7198 & 0 & 0 \\ 0 & 1.0000 & 0.1763 & 0 & 0 \\ 0 & 0 & 1.0154 & 0 & 0 \end{bmatrix} \tag{8.48}$$

$$B_{\text{lat}} = \begin{bmatrix} 0 & -1.8218 \\ 8.348 & 0 \\ 4.24 & -2.1272 \\ 0 & 0 \\ 0 & 0 \end{bmatrix} \tag{8.49}$$

$$u_{\text{lat}} = \begin{bmatrix} \delta a \\ \delta r \end{bmatrix} \tag{8.50}$$

$$x_{\text{lat}} = \begin{bmatrix} v \; p \; r \; \varphi \; \Psi \end{bmatrix}^{\mathrm{T}} \tag{8.51}$$

8.2.2 Transfer Functions

The effect of elevator and thrust inputs on the states can be represented with the transfer functions that can be derived using the Laplace transformation as:

$$\frac{Y(s)}{U(s)} = C[sI - A]^{-1}B + D \tag{8.52}$$

$$\frac{Y(s)}{U(s)} = \frac{Cadj\,(sI - A)\,B - |sI - A|}{|sI - A|} \tag{8.53}$$

where $Y(s)$ is the output and $U(s)$ is the control input. The determinant $|sI - A|$ gives the characteristic equation. MATLAB is used to calculate all the transfer functions. As an example, if the output variable is chosen as u, the forward velocity, then we can calculate $\frac{u}{\delta e}$ and $\frac{u}{\delta t}$. First, the "state_space_long = ss(Alon, Blon, [1,0,0,0,0], 0)" command gives us the state space model for the system and then the "tf(state_space_long)" command can be used to derive the equations for the forward velocity [1–3]:

$$\frac{u}{\delta e} = \frac{-0.7436s^3 - 98.75s^2 - 179.1s - 1779}{s^4 + 15.63s^3 + 88.02s^2 + 62.64s + 87.23} \tag{8.54}$$

$$\frac{u}{\delta t} = \frac{6.873s^3 + 105.1s^2 + 544.2s - 41.42}{s^4 + 15.63s^3 + 88.02s^2 + 62.64s + 87.23} \tag{8.55}$$

In a similar way, the output matrix C (given as [1,0,0,0,0] in the "ss" function) can be changed to obtain the desired transfer functions and the effect of control inputs on the rest of the longitudinal states can be found:

$$\frac{w}{\delta e} = \frac{3.785s^3 + 516.3s^2 + 179.1s + 828.5}{s^4 + 15.63s^3 + 88.02s^2 + 62.64s + 87.23}, \tag{8.56}$$

$$\frac{w}{\delta t} = \frac{-12.31s^2 - 92.89s + 8.219}{s^4 + 15.63s^3 + 88.02s^2 + 62.64s + 87.23}, \tag{8.57}$$

$$\frac{q}{\delta e} = \frac{47.92s^3 + 189.1s^2 + 168.1s}{s^4 + 15.63s^3 + 88.02s^2 + 62.64s + 87.23}, \tag{8.58}$$

$$\frac{q}{\delta t} = \frac{4.825s^2 + 68.38s}{s^4 + 15.63s^3 + 88.02s^2 + 62.64s + 87.23}, \quad (8.59)$$

$$\frac{\theta}{\delta e} = \frac{47.92s^2 + 189.1s + 168.1}{s^4 + 15.63s^3 + 88.02s^2 + 62.64s + 87.23}, \quad (8.60)$$

$$\frac{\theta}{\delta t} = \frac{4.825s + 68.38}{s^4 + 15.63s^3 + 88.02s^2 + 62.64s + 87.23}, \quad (8.61)$$

$$\frac{h}{\delta e} = \frac{-3.6s^3 + 346.5s^2 + 62.64s + 2433}{s^5 + 15.63s^4 + 88.02s^3 + 62.64s^2 + 87.23s}, \quad (8.62)$$

$$\frac{h}{\delta t} = \frac{-1.193s^3 - 6.124s^2 + 81.35s + 1106}{s^5 + 15.63s^4 + 88.02s^3 + 62.64s^2 + 87.23s}. \quad (8.63)$$

If the same process is repeated for the lateral equations of motion, we can obtain the transfer functions for the lateral motion as:

$$\frac{v}{\delta a} = \frac{-25.55s^2 + 36.49s + 195.2}{s^4 + 4.686s^3 + 26.32s^2 + 44.27s - 1.977}, \quad (8.64)$$

$$\frac{v}{\delta r} = \frac{-1.822s^2 + 35.2s + 157.5 + 154.4}{s^4 + 4.686s^3 + 26.32s^2 + 44.27s - 1.977}, \quad (8.65)$$

$$\frac{p}{\delta a} = \frac{8.348s^3 + 27.4s^2 + 211.9s - 33.36}{s^4 + 4.686s^3 + 26.32s^2 + 44.27s - 1.977}, \quad (8.66)$$

$$\frac{p}{\delta r} = \frac{10.86s^3 + 30.15s^2 + 176.5s - 27.48}{s^4 + 4.686s^3 + 26.32s^2 + 44.27s - 1.977}, \quad (8.67)$$

$$\frac{r}{\delta a} = \frac{4.24s^3 + 13.34s^2 + 47.08s + 189.2}{s^4 + 4.686s^3 + 26.32s^2 + 44.27s - 1.977}, \quad (8.68)$$

$$\frac{r}{\delta r} = \frac{-2.127s^3 - 8.454s^2 + 22.05s + 155.9}{s^4 + 4.686s^3 + 26.32s^2 + 44.27s - 1.977}, \quad (8.69)$$

$$\frac{\varphi}{\delta a} = \frac{9.096s^2 + 29.76s + 220.2}{s^4 + 4.686s^3 + 26.32s^2 + 44.27s - 1.977}, \quad (8.70)$$

$$\frac{\varphi}{\delta r} = \frac{10.48s^2 + 28.66s + 180.4}{s^4 + 4.686s^3 + 26.32s^2 + 44.27s - 1.977}, \quad (8.71)$$

$$\frac{\psi}{\delta a} = \frac{4.305s^3 + 13.55s^2 + 47.81s + 192.1}{s^5 + 4.686s^4 + 26.32s^3 + 44.27s^2 - 1.977s}, \tag{8.72}$$

$$\frac{\psi}{\delta r} = \frac{-2.16s^3 - 8.585s^2 + 22.39s + 158.3}{s^5 + 4.686s^4 + 26.32s^3 + 44.27s^2 - 1.977s}. \tag{8.73}$$

In the next sections, we investigate the characteristic equations for the stability analysis.

8.3 Longitudinal Stability Analysis

Before designing the controller, we need to investigate the stability conditions for the UAV by checking the modes for the longitudinal and lateral characteristic equations. Using the derivations given in previous sections of this chapter, the characteristic equation for the longitudinal motion can be given as follows:

$$\left(s^2 + 15.043s + 78.0719\right)\left(s^2 + 0.587s + 1.1174\right) = 0 \tag{8.74}$$

As we can see, there are two equations that describe the longitudinal motion of the UAV. These two equations represent two different modes: the phugoid mode, which is easy to control but takes a long time to dampen, and the short-period mode, which makes the plane show fluctuation motions with high frequency as the name refers and has a high damping ratio.

In the short-period mode, the changes in the attack angle and pitch angle are significant, whereas the change in the forward velocity is negligible. On the other hand, in the phugoid mode, the changes in u and pitch angle are significant, and this mode can be described as the change in the kinetic and potential energies of the aircraft. Next, we investigate the roots of the characteristic equation [2–5].

The roots of the equation are $-7.5215+4.6367i$, $-7.5215-4.6367i$, $-0.2935+1.0155i$, and $-0.2935-1.0155i$. There are two roots which are close to the imaginary axis and describe the phugoid motion and two roots which are far away from the imaginary axis and describe the short-period motion of the UAV. The angular frequency (w), damping factor (ζ), and period (T) values are calculated and given in Table 8.1. The UAV is longitudinally stable in the given condition because of the location of the roots of the characteristic equation.

Table 8.1 Short-period and phugoid mode characteristic values

Motion	Natural frequency	Damping factor	Period	Time to half amplitude
Short period	8.8538 rad/s	0.8513	1.3553 s	0.0917 s
Phugoid	1.0566 rad/s	0.0815	5.9664 s	2.3509 s

As can be seen from the calculations and the values given in the table, the roots are on the left of the imaginary axis for the longitudinal motion of the UAV and the damping factor and natural frequency values are in the desired zone.

8.4 Lateral Stability Analysis

The characteristic equation of the lateral motion is investigated and the stability condition is checked. The characteristic equation is as follows:

$$\left(s^2 + 2.5638s + 20.9734\right)(s - 0.0435)(s + 2.1658) = 0 \quad (8.75)$$

As can be seen, the lateral characteristic equation is also of the fourth order. It represents three different modes, which are the roll, Dutch roll, and spiral modes [2, 4, 6].

The roll mode is mainly a pure rolling motion. If the aircraft is not stable, this motion will lead to an increase in rolling in the direction in which the aircraft turns. The roll mode is represented by one root and gives a first-order nonoscillatory response; it is generally stable at low attack angles.

The spiral mode also gives a first-order response and it is represented by one first-order root. This mode includes both rolling and yawing motions. If it is not stable, the rolling motion increases while yawing, and the aircraft starts a spiral motion. However, instability in the spiral mode is tolerable under some specific conditions.

The Dutch roll mode is triggered by perturbations in the sideslip angle and includes yaw, roll, and sideslip changes combined. If the directional stability is not high enough to lead the aircraft into the spiral mode, the Dutch roll mode occurs and the aircraft shows oscillatory yaw and roll motions.

The roots that symbolize each motion are found from the characteristic equation and given in Table 8.2.

The results show that the spiral mode root is positive and this mode is unstable. The roll mode is stable because of the root at −2.1658. The Dutch roll mode

Table 8.2 Roll mode, Dutch roll mode and spiral mode characteristic values

Motion	Natural frequency	Damping factor	Roots	Time constant
Roll mode	N/A	N/A	−2.1658	0.4616 s
Spiral mode	N/A	N/A	0.0435 (unstable)	16 s (for double amplitude)
Dutch roll mode	4.58 rad/s	0.27	$-1.282 + 4.397i$ $-1.282 - 4.397i$	0.808 s

damping factor is 0.27 and its natural frequency is 4.58 rad/s. For stable modes, the times to get back to 36.8 % of the amplitude are given. For the spiral mode, the time to double amplitude is calculated and given in the table.

8.5 Conclusion

The linearized equations of motion of UAVs are used to find the transfer functions that relate the actuator inputs to the state outputs. The UAV is analyzed for lateral and longitudinal stability by means of these transfer functions. First, the characteristic equations are obtained using the transfer functions and the stability condition is investigated regarding the roots of the polynomials. The short-period and phugoid modes for longitudinal motion and the roll, Dutch roll, and spiral modes for lateral motion are discussed. It is shown that the UAV is stable in the analyzed flight condition.

References

1. Tewari A (2002) Modern control design with Matlab and Simulink. Wiley, Chichester
2. Pamadi BN (2003) Performance, stability, dynamics, and control of airplanes. AIAA education series, Reston
3. Etkin B, Reid LD (1996) Dynamics of flight: stability and control. Wiley, New York
4. Yechout TR, Morris SL, Bossert DE, Hallgren WF (2003) Introduction to aircraft flight mechanics. AIAA education series, Reston
5. Blakelock JH (1991) Automatic control of aircraft and missiles. Wiley, New York
6. Nelson RC (1998) Flight stability and automatic control. McGraw-Hill, New York

Chapter 9
Classic Controller Design for Unmanned Aerial Vehicles

9.1 Classical Proportional-Integral-Derivative (PID) Controller

After investigating the characteristic equations of an unmanned aerial vehicle (UAV), we can now design the control system. In this book, different methods are proposed to design a controller for UAVs; the first of these is the classical controller. In this control structure, proportional-integral-derivative (PID) type controllers are used, and longitudinal and lateral equations are investigated using the root locus method. Longitudinal and lateral controllers are designed separately.

Three different types of basic controllers and their combinations are present in the classical controller scheme. All three types have different characteristics and transfer functions. These basic controllers may be named after the type of gains used within them:

1. Proportional gain
2. Integral gain
3. Differential gain

Proportional gain has an output which is the multiple of the error between the desired and actual conditions of the state. Therefore, we can describe that:

$$c(t) = Ke(t) \tag{9.1}$$

where $c(t)$ is the output, $e(t)$ is the error for making the actual state reach the desired condition as $e(t) = x_d(t) - x_a(t)$, and K is the proportional gain. Here, $x_d(t)$ is the desired state condition and $x_a(t)$ is the actual (or observed) state.

Integral gain takes the integral of the error input and can be described by:

C. Hajiyev et al., *State Estimation and Control for Low-cost Unmanned Aerial Vehicles*, DOI 10.1007/978-3-319-16417-5_9

$$c(t) = \frac{1}{\tau_I}\int_0^t e(t)\mathrm{d}t. \tag{9.2}$$

Using the Laplace transformation, the transfer function can be represented as:

$$c(s) = \frac{1}{\tau_I s}e(s). \tag{9.3}$$

Here, τ_I is called the integral time.

Differential gain takes the derivative of the error input and can be described by the following equations:

$$c(t) = \tau_d\frac{\mathrm{d}e(t)}{\mathrm{d}t} \tag{9.4}$$

$$c(s) = \tau_d s e(s). \tag{9.5}$$

where τ_d is the differential time.

Differential gain only takes the derivative of the error and affects the future state, but cannot remove the constant error in theory. Therefore, it is not used in a control system by itself.

Different controller types can be designed using proportional, integral, and derivative gains in combination.

We can use proportional gain on its own to design a P-type controller. The system can be given by $c(t) = Ke(t)$, and its transfer function is $\frac{C(s)}{E(s)} = K$. A P-type controller normally results in a stable system. Increasing the gain value can decrease the steady-state error; however, excessive increase can also lead the system to an unstable state.

In some cases, it is possible to build a controller using only the integral gain, for which the transfer function is given as $\frac{C(s)}{E(s)} = \frac{1}{\tau_I s}$. It removes the steady-state error but can provide a very slow response.

If we integrate P- and I-type gains, what we have will be a combination of these two controllers, called a PI controller. For this type of controller, the controller input is:

$$c(t) = Ke(t) + \frac{K}{\tau_I}\int_0^t e(t)\mathrm{d}t \tag{9.6}$$

and the transfer function is:

$$\frac{C(s)}{E(s)} = K\left(1 + \frac{1}{\tau_I s}\right). \tag{9.7}$$

PI control fuses two control effects and makes the steady-state error zero. The integral action adds a root at zero to the system transfer function and removes the error at the steady state. On the other hand, the integral action also affects the overall system response, so P-type control is used to obtain the required response, compensating for the unwanted effects of the integral gain.

Another possible combination is to use the derivative effect (D) together with the proportional gain. Then, the controller is called a PD controller and its response can be given by:

$$c(t) = Ke(t) + K\tau_d \frac{\mathrm{d}e(t)}{\mathrm{d}t} \tag{9.8}$$

where the transfer function is:

$$\frac{C(s)}{E(s)} = K(1 + \tau_d s)\,. \tag{9.9}$$

In case we use the PD-type controller, the controlled system acts quickly because of the differential effect; however, the steady-state error remains. If we just want systems to reach the desired state in a short time, then this type of controller may be used in some cases.

The PID-type controller has all three control type effects and it gives the response according to the control law of:

$$C(t) = Ke(t) + \frac{K}{\tau_i}\int_0^t e(t)\mathrm{d}t + K\tau_d \frac{de(t)}{\mathrm{d}t}. \tag{9.10}$$

The transfer function can be given by:

$$\frac{C(s)}{E(s)} = K\left(\frac{1}{\tau_i s} + 1 + \tau_d s\right). \tag{9.11}$$

The PID-type controller makes the system respond quickly and removes the steady-state error. By choosing the appropriate gains, we can obtain the required response [1].

In our design, PID effects are used together to control the pitch angle, speed, heading, and altitude. A comparison is made to choose the best controller for the system. The controller coefficients are determined using the root locus method (by choosing the gain), carrying out simulations, and using optimization methods such as determining the signal response specifications through MATLAB.

9.2 Classical Controller for the Longitudinal Motion

The longitudinal controller includes a speed controller, a pitch angle controller in the inner loop, and an altitude controller in the outer loop. The scheme for the pitch angle controller is given in Fig. 9.1.

The equations for a UAV are written in the state space form and used in the blocks in MATLAB as shown in Fig. 9.2.

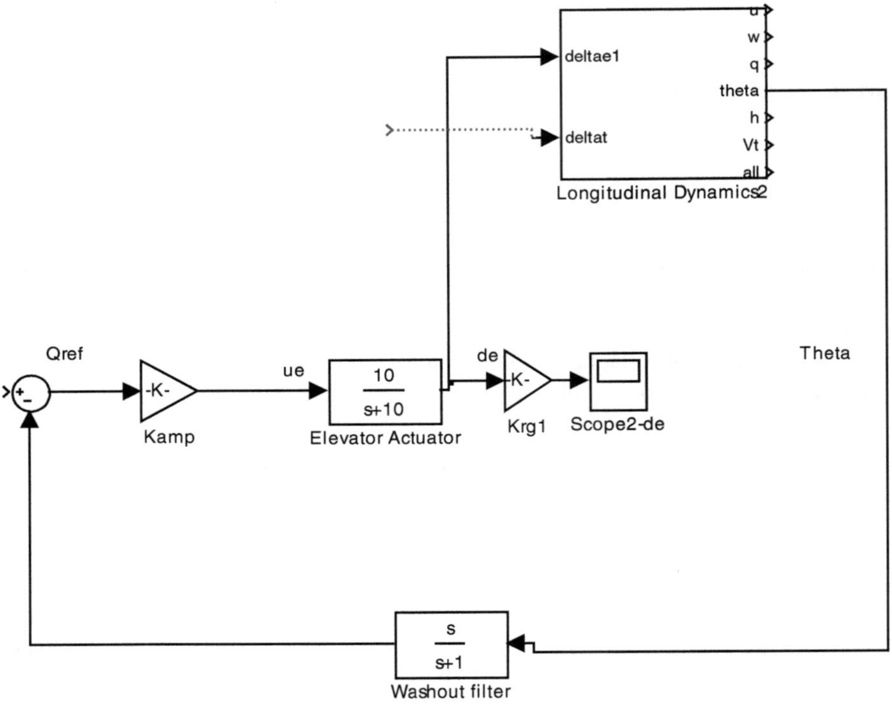

Fig. 9.1 P-type controller for pitch angle control

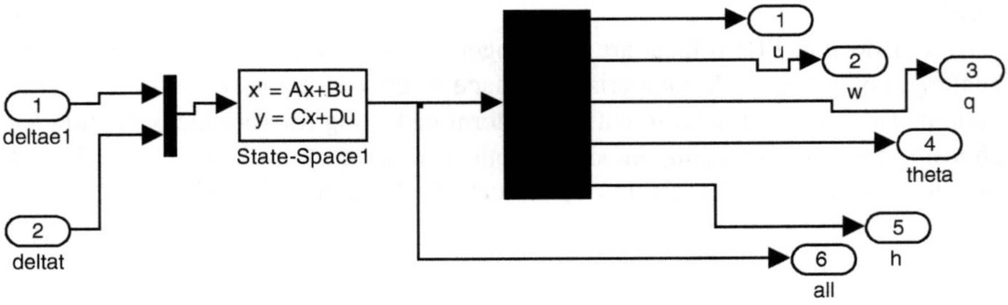

Fig. 9.2 Simulink block representing the longitudinal equations

If we use the transfer function that relates δ_e to q and include the elevator actuator function, we can choose the appropriate K_q gain for the inner loop using the root locus analysis. In the outer loop, a washout filter is added to the system to exclude the unnecessary signal inputs. For controlling θ, a P-type controller is used. K_θ can also be chosen using the root locus analysis.

9.2.1 Pitch Angular Rate Controller (Inner Loop)

The feedback for the pitch angular rate and pitch angle is usually done together to obtain the damping factor and angular frequency of the required values in the inner loop for the short-period mode and in the outer loop for the phugoid mode. However, as we can see in the linearized longitudinal equations, the values for the short-period mode are already in the required zone and feedback is not needed in the inner loop. We can check this from Military Specification MIL-F-8785C or normal aircraft standard FAR. Also, there are studies on the subject claiming that, for small UAVs, different standards should be used to evaluate the characteristics of the system [2–4]:

$$\begin{aligned} w_{1,2} &= 8,8358\text{rad/s}, \xi_1 = 0.8513, T_{1,2} = 1.3553\text{s}, t_a = 0.0917\text{rad/s} \\ w_{3,4} &= 1,0566\text{rad/s}, \xi_2 = 0.0815, T_{3,4} = 5.9664\text{s}, t_b = 2.3509\text{rad/s}. \end{aligned}$$

The short-period mode damping factor is around 0.8513 and no feedback is needed in the inner loop. Hence, instead of choosing a K_q, just the inner loop gain K_θ is used to improve the characteristic values of the phugoid mode. The inner loop may be investigated for different flight conditions and changing characteristics. The transfer function that relates θ to the u_e input can be found as:

$$\frac{\theta}{u_e} = \frac{479.2s^3 + 1,891s^2 + 1,681s}{s^6 + 26.63s^5 + 269.9s^4 + 1,187s^3 + 1,656s^2 + 1,586s + 872.3} \tag{9.12}$$

Here, the elevator actuator function, which has the time constant of 0.1 s, is modeled as:

$$TF_{\text{elevator}} = \frac{-10}{s + 10} \tag{9.13}$$

The root locus of the system for negative values of gain is shown in Fig. 9.3.

With the K_θ feedback, we can observe from the root locus that, for gain values of approximately 0.5, the short-period mode damping values do not decrease much (0.665), while the phugoid characteristics are improved (to a damping factor of 0.7). In other words, θ feedback improves the phugoid characteristics. The transfer function of the system, which includes the washout filter, and its step response is as follows (Fig. 9.4):

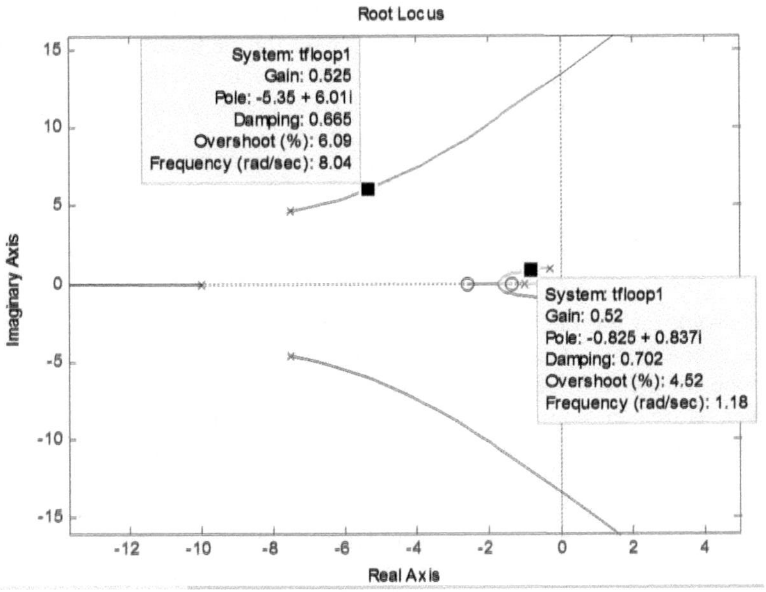

Fig. 9.3 Root locus for controlling the θ angle

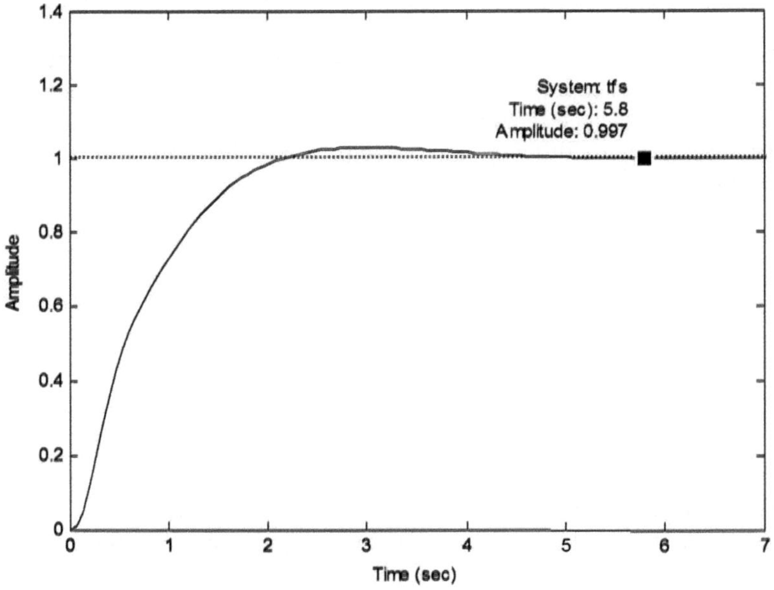

Fig. 9.4 Closed-loop response for the θ feedback loop

$$\frac{\theta}{\theta_{\text{ref}}} = \frac{L\theta}{L\theta_{\text{ref}}} \tag{9.14}$$

$$\begin{aligned} L\theta = {} & 2,492s^8 + 7,619s^7 + 94,320s^6 + 584,500s^5 + 18,160,00s^4 + 3,062,000s^3 \\ & +3,225,000s^2 + 2,244,000s^3 + 762,500 \end{aligned} \tag{9.15}$$

$$\begin{aligned} L\theta_{\text{ref}} = {} & 10^3\Big(0.001s^{11} + 0.05226s^{10} + 1.197s^9 + 15.8s^8 + 131.2s^7 + 695.4s^6 \\ & + 2,279s^5 + 4,372s^4 + 5,669s^3 + 4881s^2 + 2,768s + 760.9\Big) \end{aligned} \tag{9.16}$$

As we can see, the system is stable and the steady-state error is very low.

For the feedback loop, an equivalent feedback system transfer function can be given as:

$$TF_{\text{feedback}} = \frac{G(s)}{1 + G(s)\,(H(s) - 1)} \tag{9.17}$$

where $G(s)$ is the multiplication of the elevator actuator transfer function (θ/δ_e) and K_θ, and $H(s)$ is the washout filter transfer function [5].

The position constant where the zn are the zeros and the pn are the poles can be given as:

$$Kp = \frac{k \times zn}{pn} \tag{9.18}$$

The steady-state error can be calculated using the position constant as:

$$e_\infty = \frac{1}{1 + Kp} \tag{9.19}$$

Using the steady-state error and position constant formulas, we can choose the gain for the feedback loop to reach a low steady-state error value. In our case, the steady-state error that we obtain with the root locus-based design is 0.003, which is sufficiently small. Besides, we will build an outer loop controller for the altitude in the next section. Thus, we do not change the inner loop controller structure and we also do not use the method given by Eqs. (9.17), (9.18), and (9.19).

9.2.2 *Altitude Controller (Outer Loop)*

For the altitude controller, the performance of a PID-type controller is examined. As the first step of the design procedure, the transfer function that relates the pitch angle to the altitude is found. The transfer function from the output pitch angle to the reference input is calculated and then combined with h/θ, which can be calculated by multiplying h/δ_e and θ/δ_e. As a result, we find the transfer function of h/θ_{ref}. Then, a root locus study is made to investigate the system:

$$TF_1 = \frac{-18.72\,(s-104.5)\,(s+10)\,(s+7.34)\,(s+2.6)\,(s+1.34)\,(s+1)\,(s+0.88)}{s\,(s+12.26)\,(s+11.33)\,(s+2.6)\,(s+1.35)\,(s+0.74)\,(s^2+0.59s+1.12)} \tag{9.20}$$

$$TF_2 = \frac{\left(s^2+0.58s+1.11\right)\left(s^2+0.587+1.117\right)\left(s^2+15.04s+78\right)\left(s^2+15.04s+77.9\right)}{(s^2+0.59s+1.12)(s^2+15.04s+75)(s^2+15.04s+78)(s^2+10.7s+65)} \tag{9.21}$$

$$\left(\frac{h}{\theta_{\text{ref}}}\right) = TF_2 \times TF_1 \tag{9.22}$$

The critical gain value for stability is 0.299, as we can see in Fig. 9.5. A PID controller can be used for such a system; however, this system is already stable and it has a root at zero, which gives the integrator effect. Thus, instead of a PID controller, a PD controller is designed. The PD controller gives a quick response and is appropriate for the requirements.

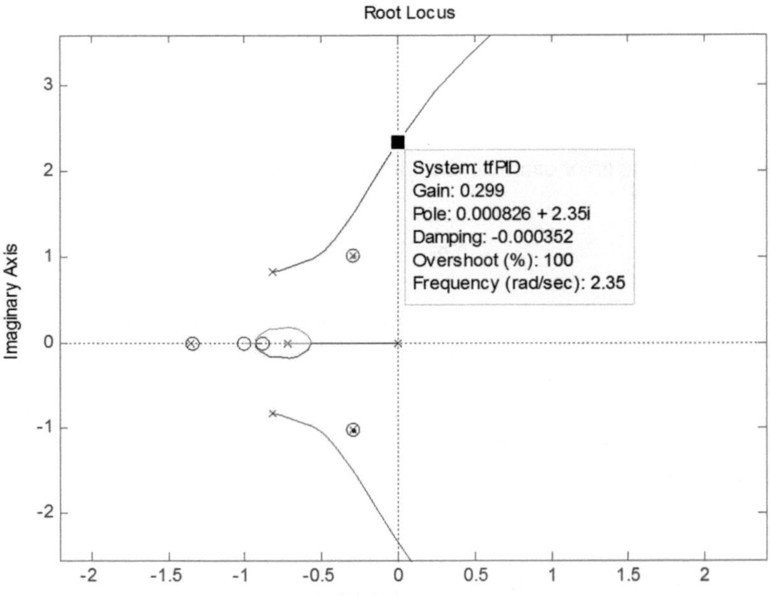

Fig. 9.5 Root locus used in the design of the altitude controller (outer loop)

Firstly, the coefficient for the PID controller, which was planned to be used, was calculated using the Ziegler–Nichols technique, such that the integral of the error is minimum [6]. The critical values of the gain and angular frequency are $K_{hpu} = 0.299$ and $\omega_u = 2.35\text{rad/s}$. According to the Ziegler–Nichols rules, the P gain must be 0.6 times the critical gain, and the integral and differential gains should be calculated accordingly [6, 7]:

$$T_u = \frac{2\pi}{\omega_u} = 2.72s \tag{9.23}$$

$$K_{hp} = 0.6 \times K_{hpu} = 0.1794 \tag{9.24}$$

$$K_{hd} = 0.6 \times K_{hpu} \times 0.125 \times T_u = 0.06 \tag{9.25}$$

$$K_{hi} = \frac{0.6 \times K_{hpu}}{0.5 \times T_{hu}} = 0.134 \tag{9.26}$$

The PID values, calculated using the Ziegler–Nichols method, are as given above. The gains are used in the system together with the saturators, which will keep the reference pitch angle at the required levels. The results are satisfactory. Both the pitch angle and the altitude reach the desired values in a short period of time.

In order to better control the speed and suppress the overshoot of the response, the gains are recalculated. The responses of the controlled and uncontrolled systems are given in Figs. 9.6, 9.7, and 9.8. Both the results found using the Ziegler–Nichols gains values and tryouts are given. The improvement in the response of the system can be seen when the gains are found by tryouts. For the tryouts, this response is obtained by taking $K_{hp} = 0.58$, $K_{hi} = 0.25$, and $K_{hi} = 0.2$.

The response of the altitude system is improved. However, the integral effect is already contained in the system equations, and we prefer to use a PD-type controller, not a PID-type controller. The gains of the PD-type controller are determined using the Ziegler–Nichols values as a starting point. The proportional gain value K_{hp} is chosen as 0.1794, and tryouts are done using the previously determined K_{hd}. The final value of K_{hd} that is used in the PD controller is 0.132. In this case, a zero at -1.33 is added to the system and a nonvibratory quick response is obtained. The step response of the system with the PD controller is given in Fig. 9.9.

The response of the designed altitude controller for the case where the reference input is 100 m and the speed is held constant at 20 m/s using the speed controller is given in Fig. 9.10. Moreover, the response of the speed controller, which tries to hold the speed at 20 m/s, is given in Fig. 9.11. As we can see, the PD controller is working efficiently as an altitude controller for our system. Nonetheless, in cases where other disturbances are present, a PID-type controller may be used instead.

In Fig. 9.12, the Simulink diagram for the final structure of the altitude controller is given. As can be seen, the control is obtained using the K_θ in the inner loop for pitch angle control and the PD-type controller in the outer loop for altitude control

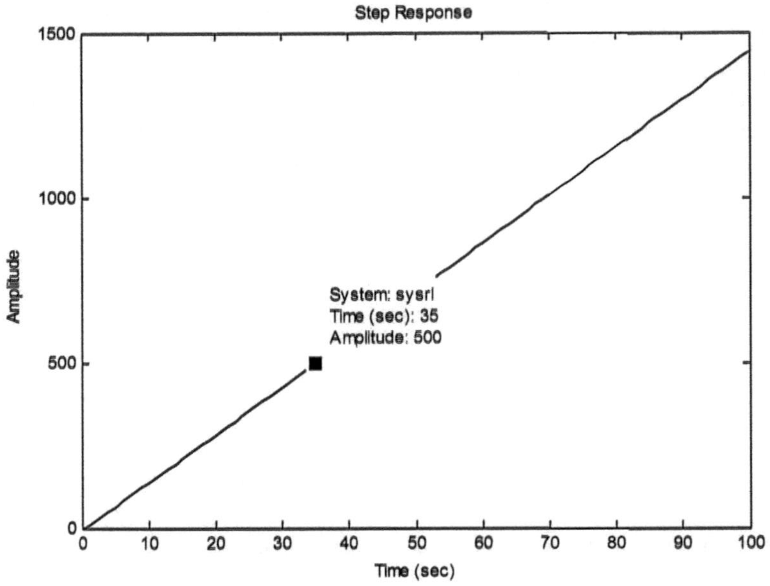

Fig. 9.6 Response of the uncontrolled system (h/θ_{ref})

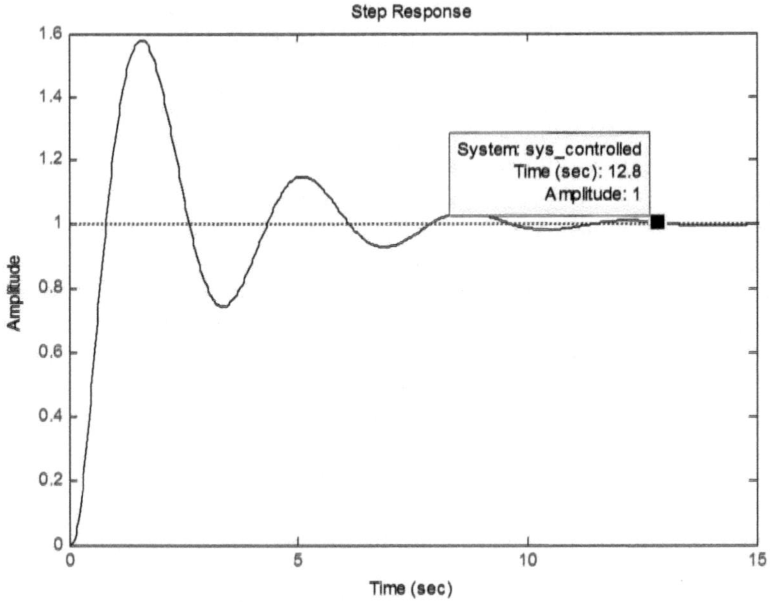

Fig. 9.7 Response of the altitude control system with a PID controller using Ziegler–Nichols gains

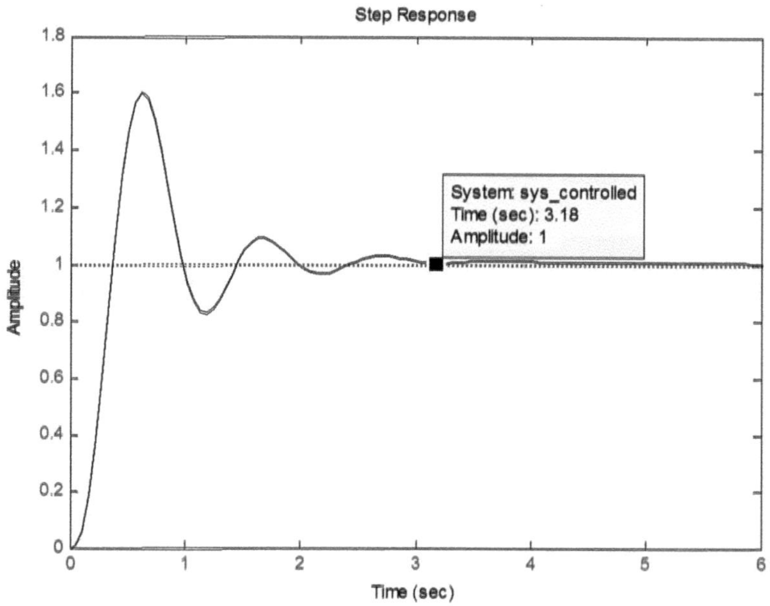

Fig. 9.8 Response of the altitude control system with a PID controller using the gains found in tryouts

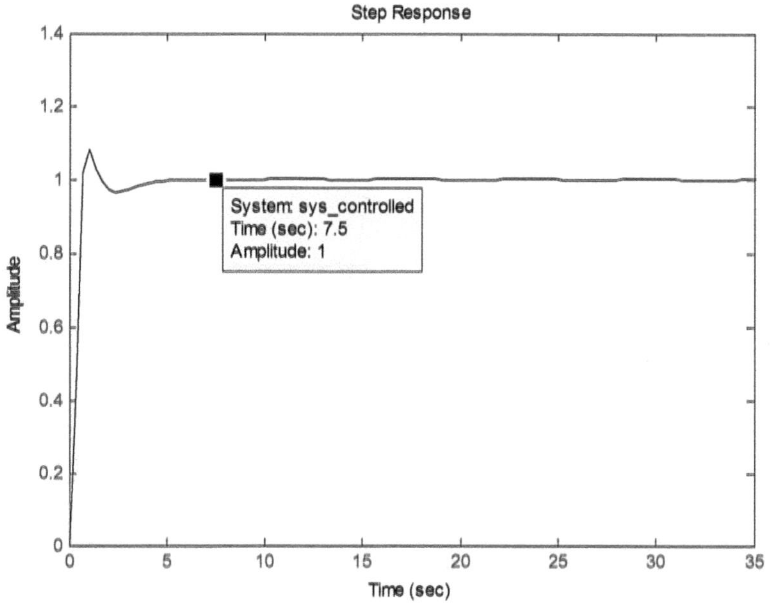

Fig. 9.9 Step response of the PD controller

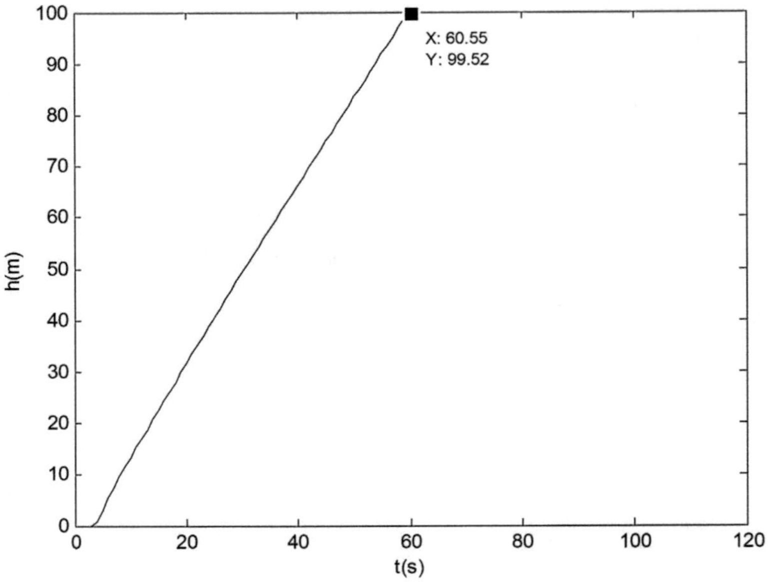

Fig. 9.10 Response of the PD-type altitude controller to a 100-m reference input

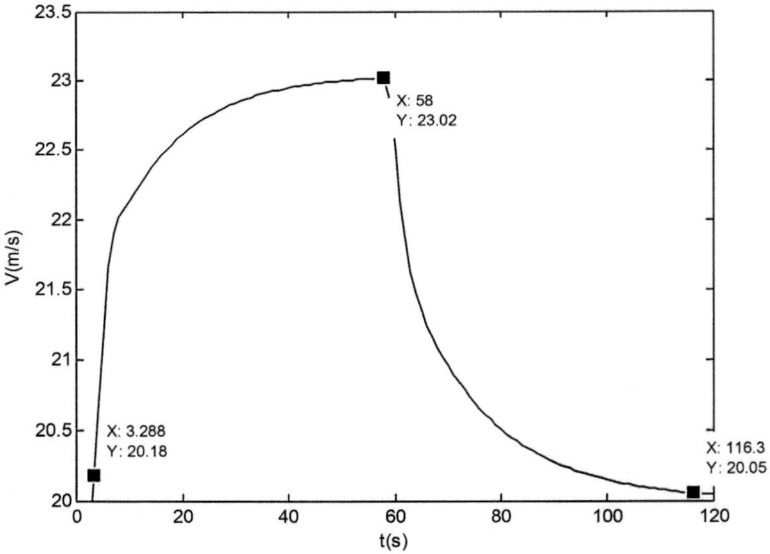

Fig. 9.11 Response of the system when the speed is trying to be held at 20 m/s (while the altitude controller is working with the reference input)

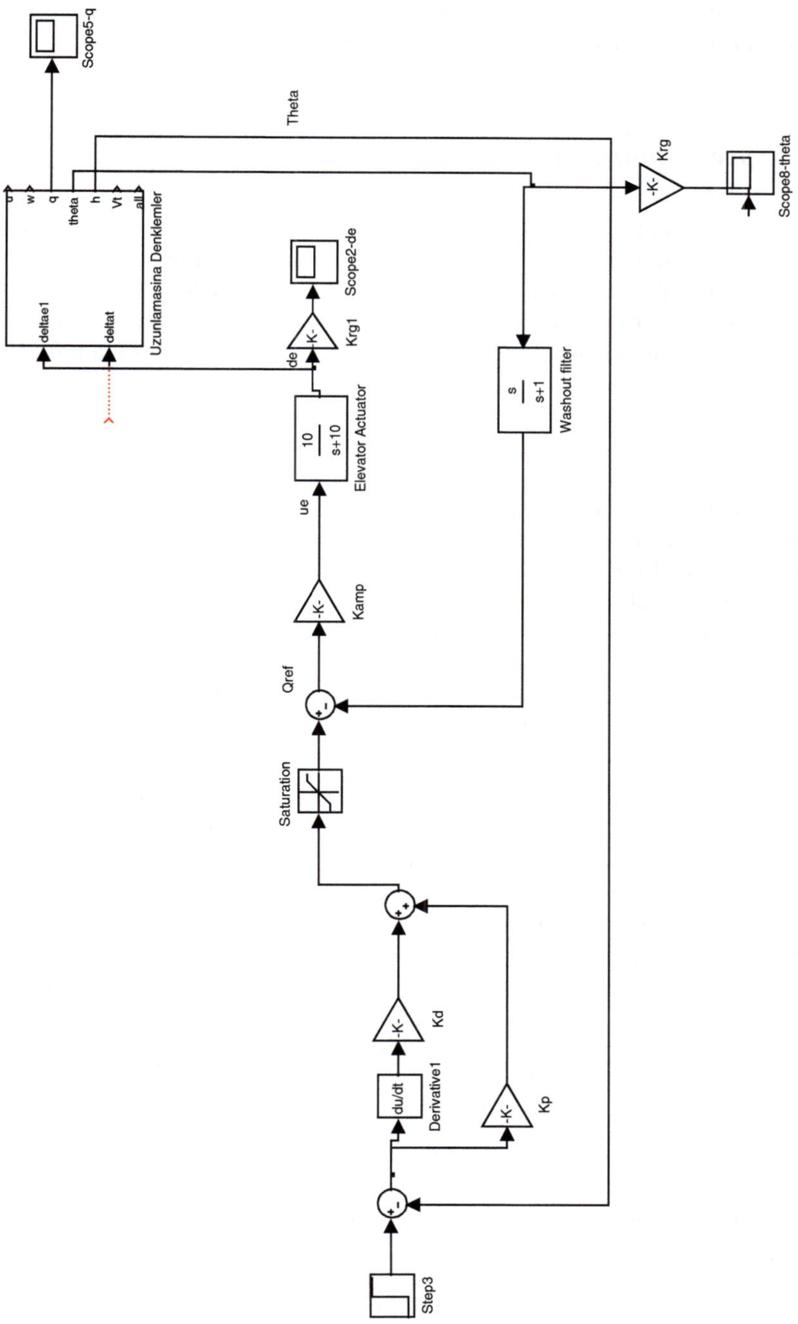

Fig. 9.12 Simulink diagram for the altitude controller

and saturators. The saturators are used for limiting the elevator angle and reference pitch angle values. Although not shown in the diagram, the δt input is determined using the speed controller.

9.2.3 Speed Controller

A controller that controls the speed using the thrust input, δt, is designed. A P-type controller is used. Both the results of the step response and the simulations show that a P-type speed controller which uses δt as the input is satisfactory. In order to find the gain value K, the root locus method is used. The $V_t/\delta t$ transfer function does not normally exist in the previously found functions; however, we can obtain it using the equation below:

$$\dot{V}_t V_t = U\dot{U} + W\dot{W} + V\dot{V} \tag{9.27}$$

We can find the $V_t/\delta t$ transfer function if we assume that, for a given linearization point, $\frac{u}{V_t}$ and $\frac{w}{V_t}$ are almost constant and v is zero. The other values can be determined using the calculated values for the speed values, etc.

A sixth row to include the V_t transfer function is added to the system. The new A_{lon} matrix, denoted $A_{\text{lon\_vt}}$, can be given as:

$$A_{\text{lon\_vt}} = \begin{bmatrix} -0.3356 & 1.3181 & -1.9276 & -9.6610 & 0 & 0 \\ -1.7916 & -3.9003 & 9.8215 & -1.7035 & 0 & 0 \\ 0.7020 & -3.5375 & -11.3920 & 0 & 0 & 0 \\ 0 & 0 & 1.0000 & 0 & 0 & 0 \\ -0.1736 & -0.9848 & 0 & 17.4865 & 0 & 0 \\ -1.234 & -3.5261 & 9.1667 & 3.8967 & 0 & 0 \end{bmatrix} \tag{9.28}$$

As explained previously, the measured value can be chosen as V_t and then the related transfer function can be calculated. The transfer function for the thrust actuator is modeled with a first-order transfer function with a time constant of 0.5 s:

$$\frac{V_t}{\delta t} = \frac{1.146s^4 + 9.433s^3 + 58.81s^2 + 318.8s + 423.2}{s^5 + 15.63s^4 + 88.02s^3 + 62.64s^2 + 87.24s} \tag{9.29}$$

$$\text{Actuator}TF = \frac{2}{s+2} \tag{9.30}$$

The open-loop transfer function can be obtained by combining these two transfer functions:

$$\frac{V_t}{u_t} = \frac{2.292s^4 + 18.87s^3 + 117.6s^2 + 637.6s + 846.4}{s^6 + 17.63s^5 + 119.3s^4 + 238.7s^3 + 212.5s^2 + 174.5s} \tag{9.31}$$

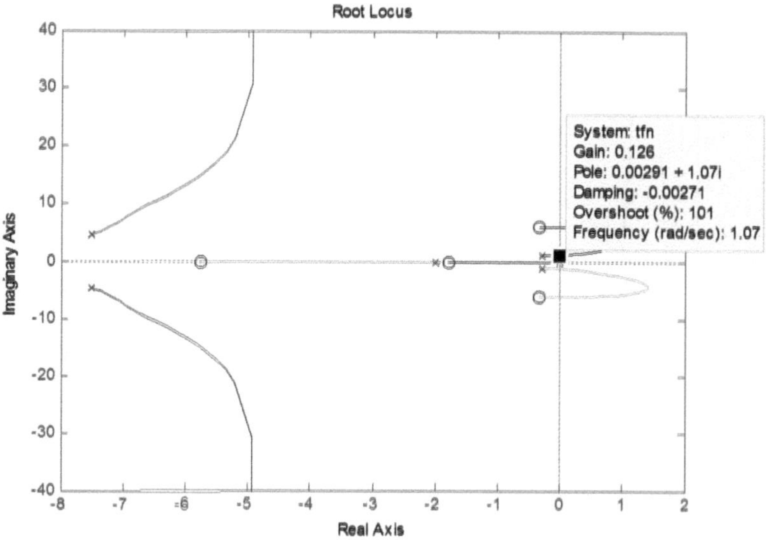

Fig. 9.13 The root locus diagram for $V_t/\delta t$

The root locus for the obtained $V_t/\delta t$ transfer function is as in Fig. 9.13. As can be seen, the system is stable for gains smaller than 0.126 ($K < 0.126$). Firstly, the PID gains proposed by the Ziegler method are checked. The determined critical values are $K_{pu} = 0.126$ and $\omega_u = 1.07$ rad/s. The calculation for the gains are as follows:

$$T_u = \frac{2\pi}{\omega_u} = 5.8721s, \tag{9.32}$$

$$K_p = 0.6 \times K_{pu} = 0.6 \times 0.126 = 0.0756, \tag{9.33}$$

$$K_d = 0.6 \times K_{pu} \times 0.125 \times T_u = 0.0257, \tag{9.34}$$

$$K_i = \frac{0.6 \times K_{pu}}{0.5 \times T_u} = 0.0555. \tag{9.35}$$

In Fig. 9.14, we see that the overshoot values and settling time are not good. Hence, we designed a simpler controller using only the proportional gain value. The system is already stable because of the roots of the transfer function. The response of the system when $K_{vt} = 0.013$ is given in Fig. 9.15. We may obtain better responses by increasing the gain value; however, the change in δt to obtain the better responses affects the overall system response. The response of the system when the gain value is chosen as 0.025 is given in Fig. 9.16.

It is clear that the response quickens. However, the response of the overall control system is also affected. In Figs. 9.17 and 9.18, the results for the altitude change

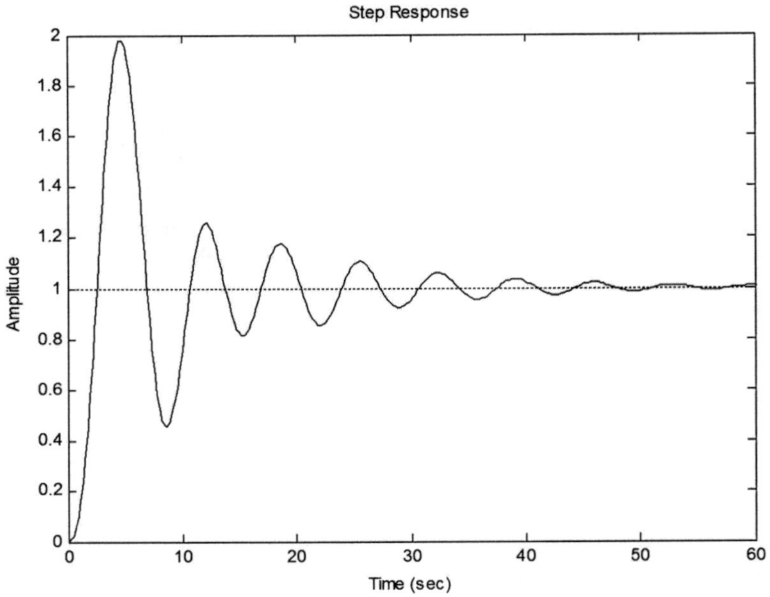

Fig. 9.14 PID controller response for the gains calculated using the Ziegler–Nichols method

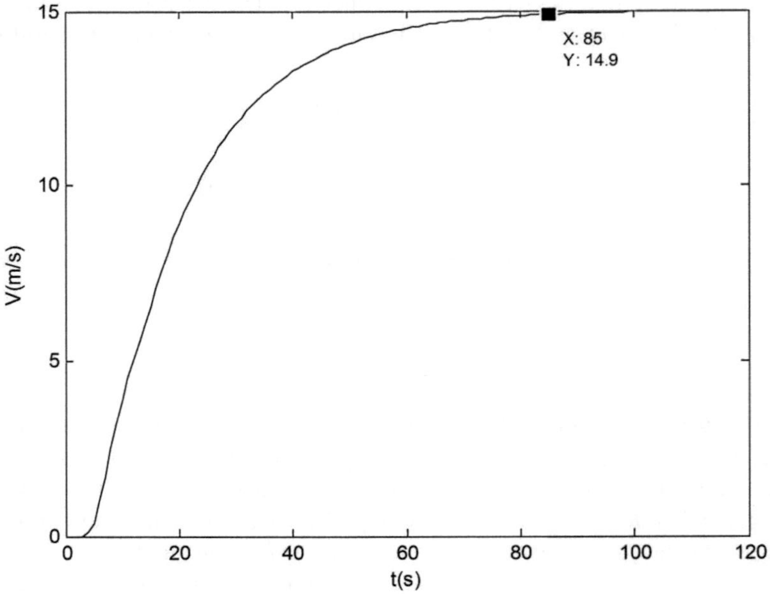

Fig. 9.15 Response of the speed controller system to a 15-m/s reference input with only P gain ($K_{vt} = 0.013$)

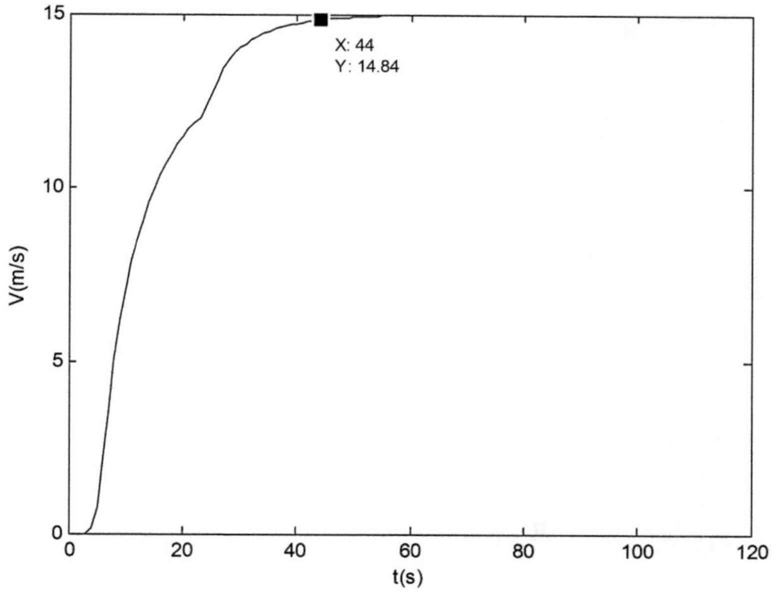

Fig. 9.16 Response of the speed controller system to a 15-m/s reference input with only P gain ($K_{vt} = 0.025$)

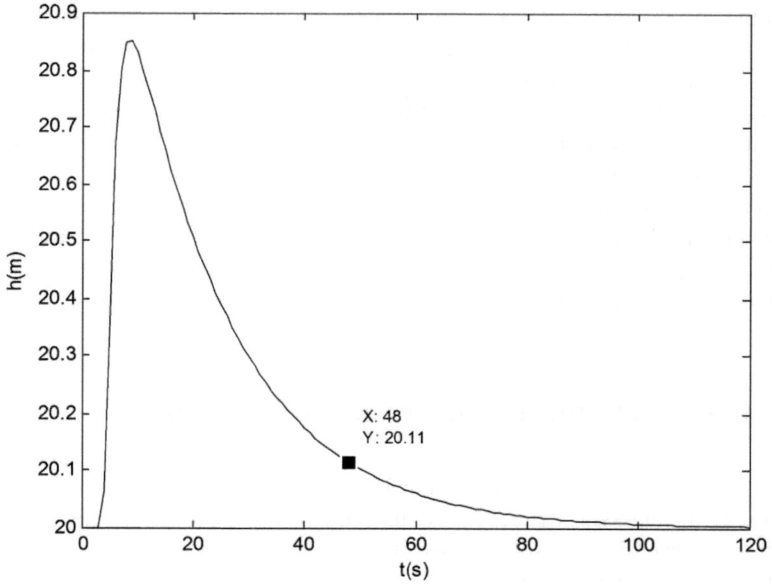

Fig. 9.17 Change in altitude when the speed controller changes the UAV's speed by 15 m/s ($K_{vt} = 0.015$)

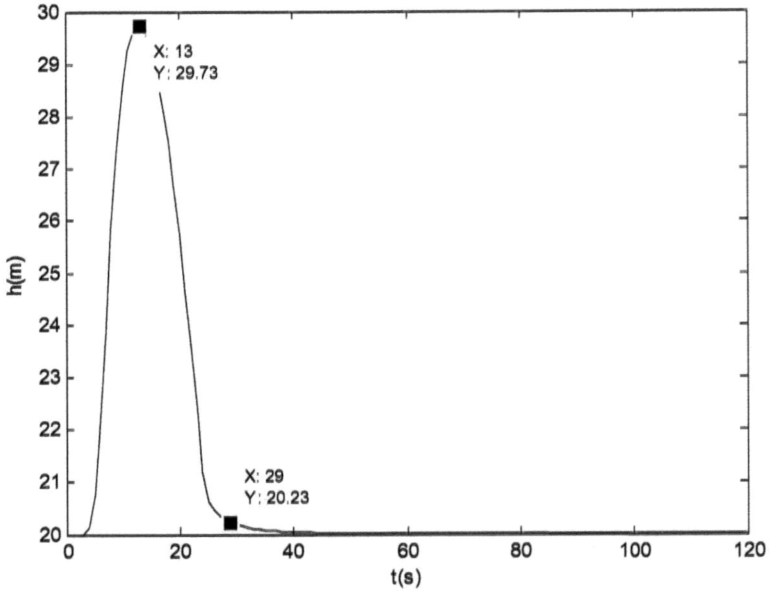

Fig. 9.18 Change in altitude when the speed controller changes the UAV's speed by 15 m/s ($K_{vt} = 0.025$)

when the speed value is set to a certain value are given. It can be understood that, when the gain used for controlling V_t is increased, the change in altitude is also increased. Changes of around 10 m in altitude over a short period, as we see for the case $K_{vt} = 0.025$, is an undesired response for mini-UAV operations (Fig. 9.18). Thus, a gain value that gives a quick response for controlling V_t and results in lower altitude changes must be used.

In Fig. 9.19, the step response for the controllerwith $K_{vt} = 0.015$ is given. This is the final gain value that we choose for the speed controller. For speed changes below 10 m/s, it may be increased to $K_{vt} = 0.02$, depending on the desired response.

The overall structure for the speed controller is given in Fig. 9.20.

9.3 Classical Controller for the Lateral Motion

The equations that we derived for the lateral motion were given in Chap. 8. In addition to the classical states, we also used the heading equation, which was given as $\dot{\psi} = r \sec\theta$. In this section, we design a control system using the derived equations and the matrices A_{lat} and B_{lat}.

For the lateral classical controller design, we prefer a method that consists of inner loops for increasing the stability. A yaw damper is used to reduce the rudder

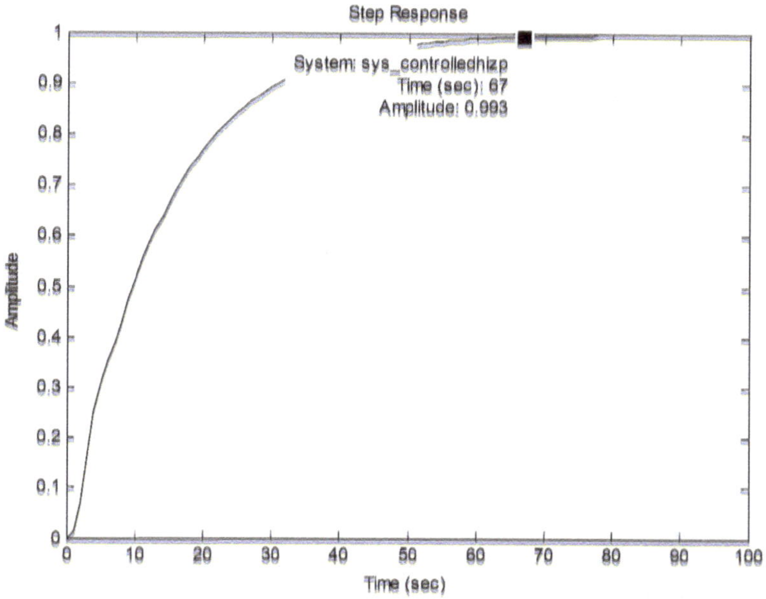

Fig. 9.19 Step response of the system for $K_{vt} = 0.015$

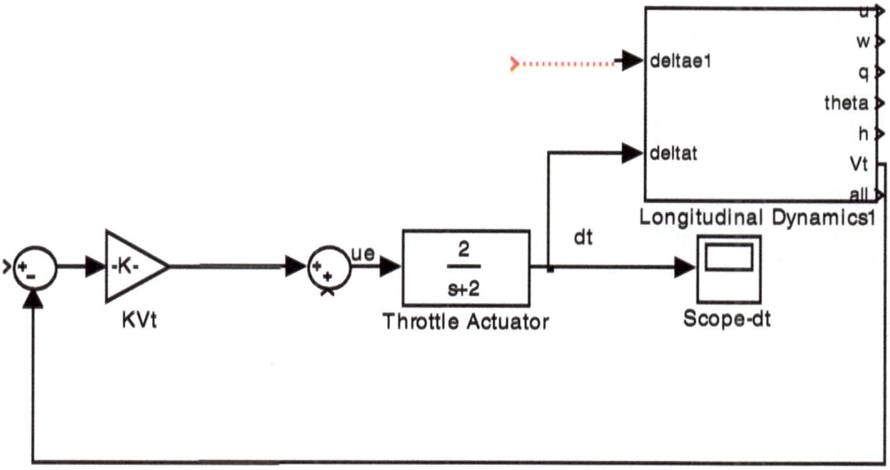

Fig. 9.20 Diagram of the speed controller

reference value to zero when the controller for the heading is working. Changes in the roll rate (p), roll angle (Φ), and heading angle (Ψ) are used to determine the required input to the aileron. This method is a classical and effective method for designing the lateral controller. However, during the design process, the coupling

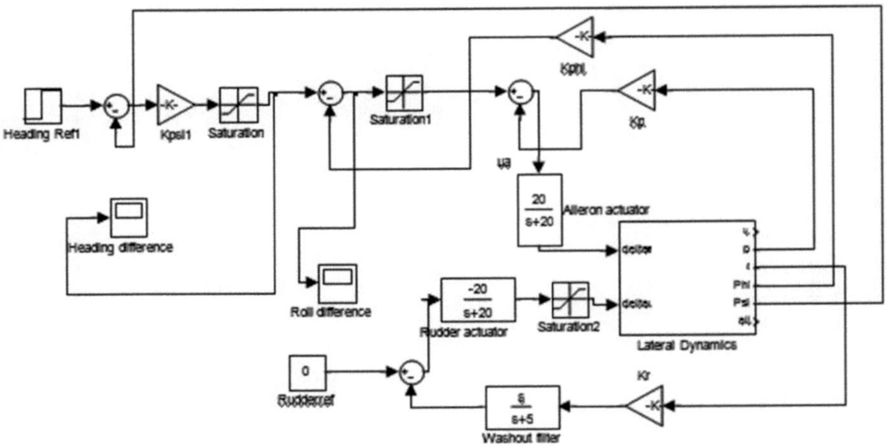

Fig. 9.21 Lateral control system scheme

between the aileron and rudder should be kept in mind to overcome the difficulties; washout filter, aileron and rudder equations should be handled together, and actuator equations should be included in the state equations while the controller is designed (Fig. 9.21).

9.3.1 Roll Rate Controller

We can start the design with the p loop that has the least effect on the controller [8]. p/δ_a can be obtained using the lateral equations. The aileron actuator transfer function is chosen as $\frac{20}{20+s}$.

$$\frac{p}{\delta_a} = \frac{8.348s^3 + 27.4s^2 + 211.9s - 33.36}{s^4 + 4.686s^3 + 26.32s^2 + 44.27s - 1.977} \quad (9.36)$$

Combining the given transfer functions, we can find p/u_a (roll angle change/control input):

$$\frac{p}{u_a} = \frac{167s^3 + 548s^2 + 4,238s - 667.2}{s^5 + 24.69s^4 + 120s^3 + 570.7s^2 + 883.4s - 39.54} \quad (9.37)$$

The root locus drawn for this system is given in Fig. 9.22.

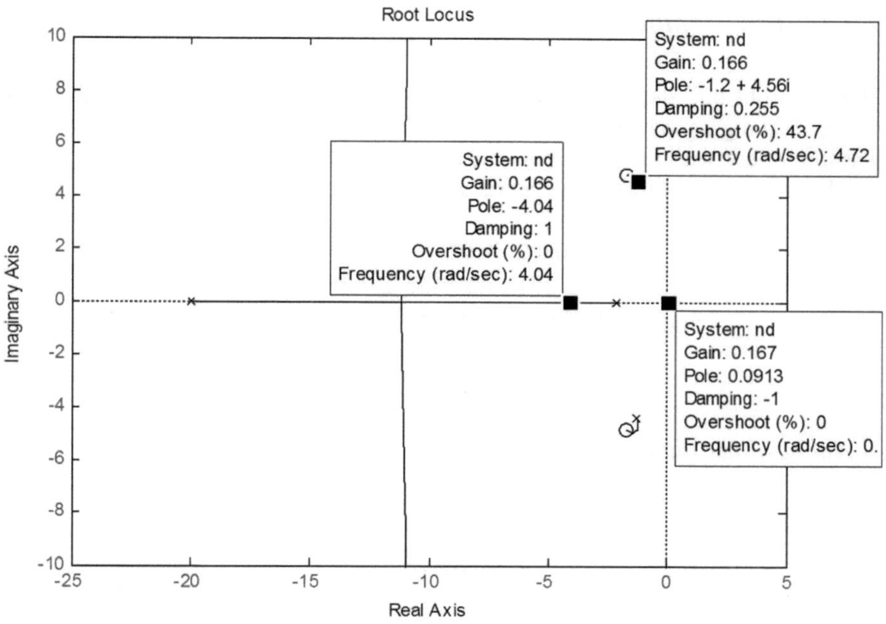

Fig. 9.22 Root locus diagram for p/u_a

What we see from the root locus is that the root of the roll mode equation must be moved to the left on the real axis to increase the stability of the system and the damping factor of roll angle change. The roots of the characteristic equation previously determined are:

$$s_{\text{spiral}} = 0.0435, s_{\text{roll}} = -2.1658, s_{\text{dutchroll}} = -1.2819 + 4.3966i, -1.2819 - 4.3966i.$$

From the root locus, we understand that the root may be moved to −4.03 by choosing a gain value of 0.165. At the same time, this gain value faintly decreases the Dutch roll mode damping factor. We also need to check the value of the spiral mode root. The spiral mode root is unstable; however, when the gain is changed, the root becomes 0.09 and the time for double amplitude can be found as:

$$T = \frac{\log(2)}{0.09} = 7.5s, \tag{9.38}$$

which is good enough to ensure a normal working condition. Thus, the gain value K_p can be chosen as 0.165.

The second loop to be closed is r/δ_r (change in yaw angle by rudder input).

9.3.2 Yaw Damper

Starting from this step of the lateral controller design, we need to define output matrix C and gain matrix K by taking the coupling effect in lateral motion and the washout filter used in the r/δ_r loop into account.

The state matrix (A_{lat}) of the lateral system is:

$$A_{\text{lat}} = \begin{bmatrix} -1.0502 & 1.9276 & -9.8215 & 9.6610 & 0 \\ -1.2213 & -1.9155 & 1.0096 & 0 & 0 \\ 1.7255 & 0.0919 & -1.7198 & 0 & 0 \\ 0 & 1.0000 & 0.1763 & 0 & 0 \\ 0 & 0 & 1.0154 & 0 & 0 \end{bmatrix} \tag{9.39}$$

The new state matrix can be found as:

$$X = \begin{bmatrix} \text{A}_{\text{lat}} & \text{B}_{\text{lat}} & 0 \\ 0\,0\,0\,0\,0 - 1/T_{\text{a}}\,0\,0 \\ 0\,0\,0\,0\,0\,0 - 1/T_{\text{r}}\ 0; \\ 0\,0\,1\,0\,0\,00 - 1/T_{\text{w}} \end{bmatrix} \tag{9.40}$$

Here, δ_a, aileron input, δ_r, rudder input, and x_w, washout, are added into the lateral state matrix. T_{a}, T_{r}, and T_{w} are the time constants for the aileron actuator, rudder actuator, and washout filter, respectively. The new lateral state matrix is:

$$X = \begin{bmatrix} -1.0502 & 1.9276 & -9.8215 & 9.6610 & 0 & 0 & -1.8218 & 0 \\ -1.2213 & -1.9155 & 1.0096 & 0 & 0 & 8.3479 & 10.8560 & 0 \\ 1.7255 & 0.0919 & -1.7198 & 0 & 0 & 4.2400 & -2.1272 & 0 \\ 0 & 1.0000 & 0.1763 & 0 & 0 & 0 & 0 & 0 \\ 0 & 0 & 1.0154 & 0 & 0 & 0 & 0 & 0 \\ 0 & 0 & 0 & 0 & 0 & -20.0000 & 0 & 0 \\ 0 & 0 & 0 & 0 & 0 & 0 & -20.0000 & 0 \\ 0 & 0 & 1.0000 & 0 & 0 & 0 & 0 & -5 \end{bmatrix} \tag{9.41}$$

In this case, the control distribution matrix B becomes:

$$B = \begin{bmatrix} 0 & 0 \\ 0 & 0 \\ 0 & 0 \\ 0 & 0 \\ 20 & 0 \\ 0 & 20 \\ 0 & 0 \end{bmatrix} \tag{9.42}$$

The values in the new B matrix are the rudder and aileron actuator constants. The control inputs and states in this case are:

$$u_{\text{lat}} = \begin{bmatrix} u_a \\ u_r \end{bmatrix} \tag{9.43}$$

$$x_{\text{lat}} = \begin{bmatrix} v \; p \; r \; \varphi \; \Psi \; \delta_a \; \delta_r \; x_w \end{bmatrix}^{\text{T}} \tag{9.44}$$

At this point, if we can define the necessary C and K matrices, then the feedback can be achieved by calculating:

$$X_{\text{lat}} = X - BKC \tag{9.45}$$

The matrix C includes feedback from x_w, p, and φ. In B, one input from the aileron and one input from the rudder are present. K must be determined at each step using the appropriate feedback channel.

For example, for the first step, the gain matrix is:

$$K = \begin{bmatrix} 0.1650 & 0 & 0 \\ 0 & 0 & 0 \end{bmatrix}, \tag{9.46}$$

and C can be determined as:

$$C = \begin{bmatrix} 0 & 1 & 0 & 0 & 0 & 0 & 0 & 0 \\ 0 & 0 & 1 & 0 & 0 & 0 & 0 & -5 \\ 0 & 0 & 0 & 1 & 0 & 0 & 0 & 0 \end{bmatrix}. \tag{9.47}$$

where x_w, p, and φ are chosen as the outputs. In this way, for the first step, the following matrix can be calculated:

$$BKC = \begin{bmatrix} 0 & 0 & 0 & 0 & 0 & 0 & 0 & 0 \\ 0 & 0 & 0 & 0 & 0 & 0 & 0 & 0 \\ 0 & 0 & 0 & 0 & 0 & 0 & 0 & 0 \\ 0 & 0 & 0 & 0 & 0 & 0 & 0 & 0 \\ 0 & 0 & 0 & 0 & 0 & 0 & 0 & 0 \\ 0 & 3.3000 & 0 & 0 & 0 & 0 & 0 & 0 \\ 0 & 0 & 0 & 0 & 0 & 0 & 0 & 0 \\ 0 & 0 & 0 & 0 & 0 & 0 & 0 & 0 \end{bmatrix} \tag{9.48}$$

The system must be updated using the newly calculated A matrix and finding r/δ_r by means of the x_w state. The new lateral state matrix with the first feedback is:

$$X1 = \begin{bmatrix} -1.0502 & 1.9276 & -9.8215 & 9.6610 & 0 & 0 & -1.8218 & 0 \\ -1.2213 & -1.9155 & 1.0096 & 0 & 0 & 8.3479 & 10.8560 & 0 \\ 1.7255 & 0.0919 & -1.7198 & 0 & 0 & 4.2400 & -2.1272 & 0 \\ 0 & 1.0000 & 0.1763 & 0 & 0 & 0 & 0 & 0 \\ 0 & 0 & 1.0154 & 0 & 0 & 0 & 0 & 0 \\ 0 & -3.3000 & 0 & 0 & 0 & -20.0000 & 0 & 0 \\ 0 & 0 & 0 & 0 & 0 & 0 & -20.0000 & 0 \\ 0 & 0 & 1.0000 & 0 & 0 & 0 & 0 & -5 \end{bmatrix} \quad (9.49)$$

Now, we can find the required transfer functions as:

$$\frac{r}{\delta_r} = \frac{-42.54s^4 - 1,020s^3 - 7,151s^2 + 5,162s + 6,2350}{s^6 + 44.69s^5 + 641.3s^4 + 3,613s^3 + 14,810s^2 + 3,1510s - 2,992}, \quad (9.50)$$

$$\frac{r}{u_r} = \frac{-42.54s^5 - 1,020s^4 - 7,151s^3 + 5,162s^2 - 6,2350s}{s^7 + 49.69s^6 + 864.8s^5 + 6,820s^4 + 32,875s^3 + 105,560s^2 + 154,558s - 14,960}. \quad (9.51)$$

r/u_r is calculated by adding the washout filter transfer function to r/δ_r.

The roots of the characteristic equation can be calculated as -20.0642, -18.2896, $-1.1988 + 4.5611i$, $-1.1988 - 4.5611i$, -5.0000, -4.0297, and 0.0910.

The purpose of using feedback in this loop is to increase the Dutch roll mode damping factor. As can be understood from the root locus graphic (Fig. 9.23), for

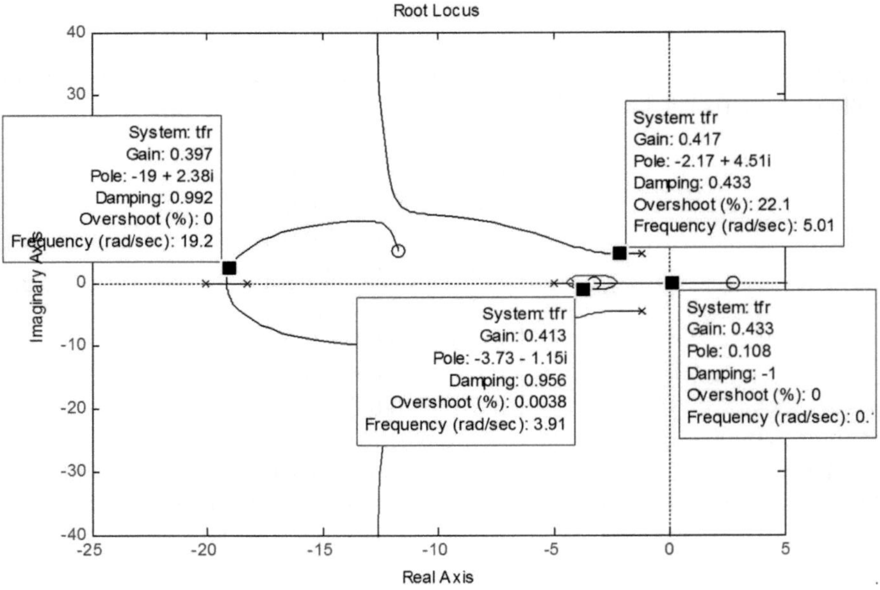

Fig. 9.23 Root locus diagram for the yaw angle loop

negative gains, a gain value of 0.405 makes the damping factor 0.43 and does not affect the spiral mode root much. Therefore, K_r can be chosen as −0.4. After this step, we must find φ/δ_a, which is the equation that is necessary for φ feedback.

9.3.3 Roll Angle Loop

Similar steps to those introduced in the previous section can be used to find the equation that relates φ to δ_a and u_a:

$$\frac{\varphi}{u_a} = \frac{181.9s^4 + 5,143s^3 + 4,768s^2 + 186,000s + 440,400}{s^7 + 49.69s^6 + 881.7s^5 + 7,227s^4 + 35,730s^3 + 103,500s^2 + 129,600s - 14,960} \tag{9.52}$$

The roots of the characteristic equation are: $-19.0107 + 2.3780i$, $-19.0107 - 2.3780i$, $-2.1175 + 4.5109i$, $-2.1175 - 4.5109i$, $-3.7698 + 1.1211i$, $-3.7698 - 1.1211i$, and 0.1061.

Gain values are determined using the root locus diagram (Fig. 9.24).

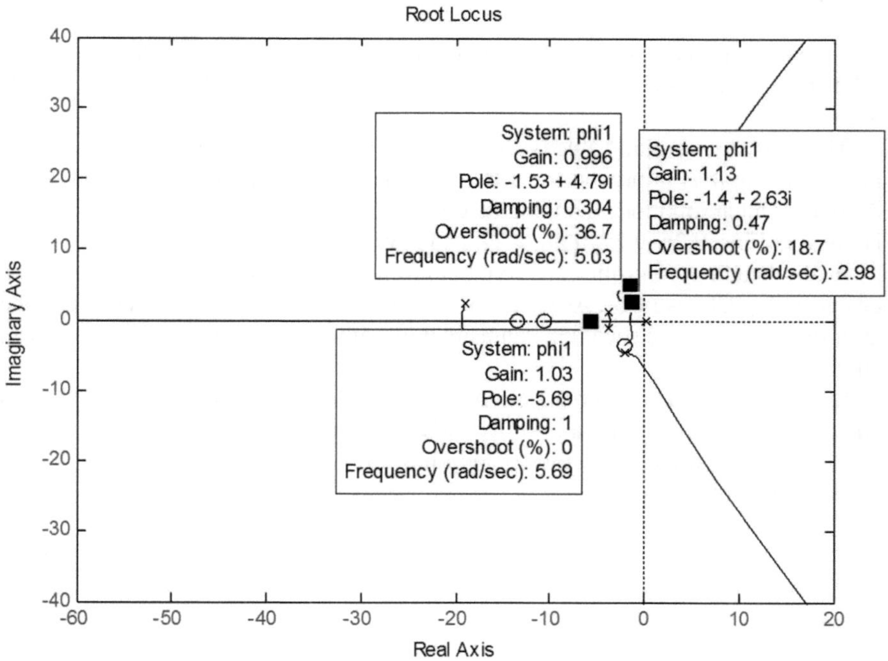

Fig. 9.24 Root locus diagram for the roll angle feedback loop

A gain value of $K_\varphi \approx 1$ means that the spiral mode root becomes more stable and the Dutch roll mode damping factor also stays high. In this case, the roll mode root stays at a value of 5.5. This choice of K gives us a nonvibratory response. Also, choosing a high proportional gain value decreases the steady-state error value.

9.3.4 Heading Controller

In order to design the heading controller, we should use a last gain matrix to determine the equation that relates heading to the aileron input. The last gain matrix we found is:

$$K_{\text{last}} = \begin{bmatrix} 0.1650 & 0 & 1 \\ 0 & -0.4 & 0 \end{bmatrix} \tag{9.53}$$

Recalculating the lateral A matrix and choosing the appropriate output, we can find:

$$\frac{\psi}{u_a} = \frac{86.11s^5+2{,}424s^4+16{,}340s^3+54{,}840s^2+191{,}700s+384{,}200}{s^8+49.69s^7+881.7s^6+7{,}409s^5+40{,}870s^4+151{,}100s^3+315{,}600s^2+425{,}400s} \tag{9.54}$$

The open-loop roots of the last loop are: $-19.117+1.92i$, $-19.117-1.92i$, $-1.5285+4.79i$, $-1.5285-4.79i$, -5.6534, $-1.3728+2.4857i$, and $-1.3728-2.4857i$.

The proportional gain value is chosen as 0.5 (Fig. 9.25). With this value, the roll mode root value is improved for better response and the spiral mode is stabilized. Also, the Dutch roll mode damping factor becomes 0.3. The response of the system is given in Figs. 9.26, 9.27, 9.28, 9.29, and 9.30.

In the heading controller, a P-type controller is chosen because of its simple design and quick response. As we can see from the figures, the controller directs the aircraft to the reference heading value quickly, without significant change in the roll angle.

In this system, the rudder and aileron actuator time constants are taken as 0.05 s and their transfer functions are positive and negative $\frac{20}{s+20}$, respectively.

The washout filter in the rudder loop has a time constant of $T=0.2$ s, and its transfer function is $\frac{s}{s+5}$.

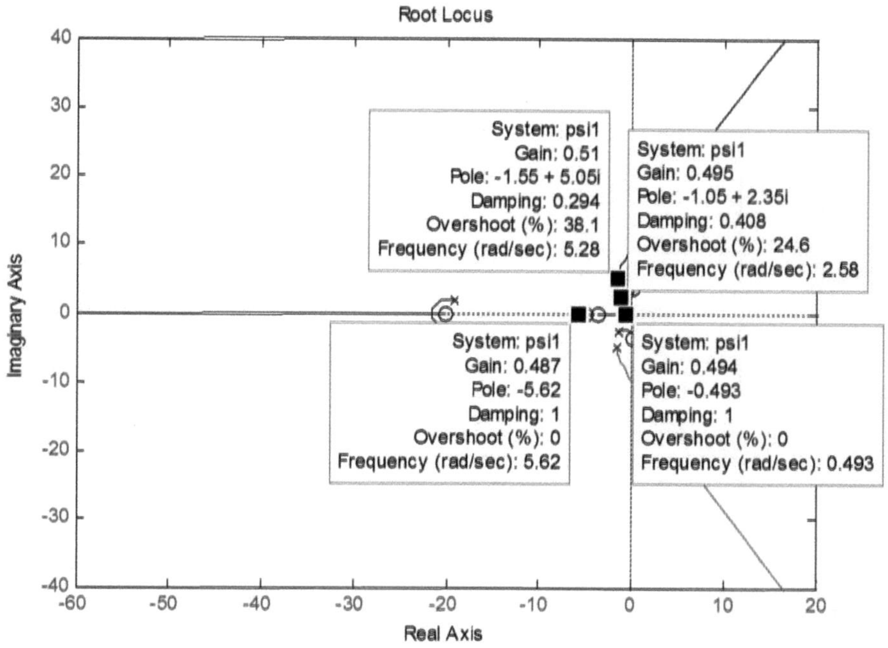

Fig. 9.25 Root locus value for $K_\Psi = 0.5$

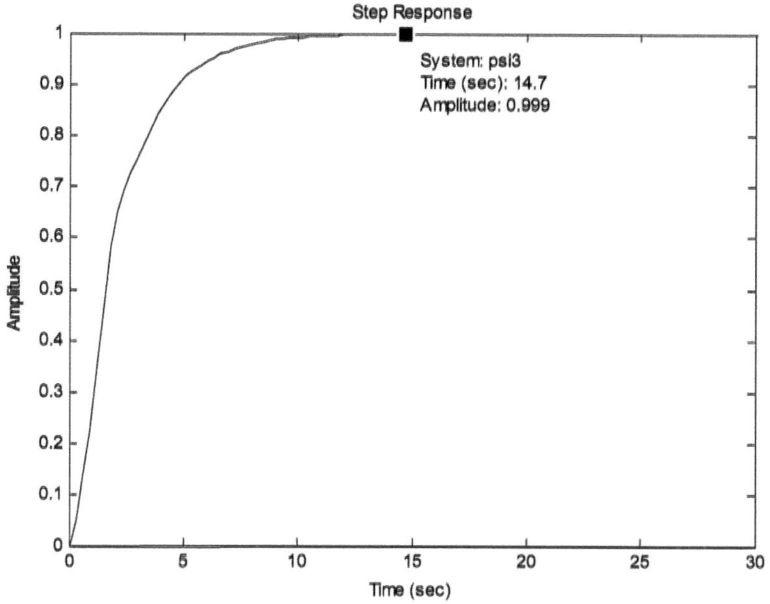

Fig. 9.26 Step response of the system to unit heading reference input

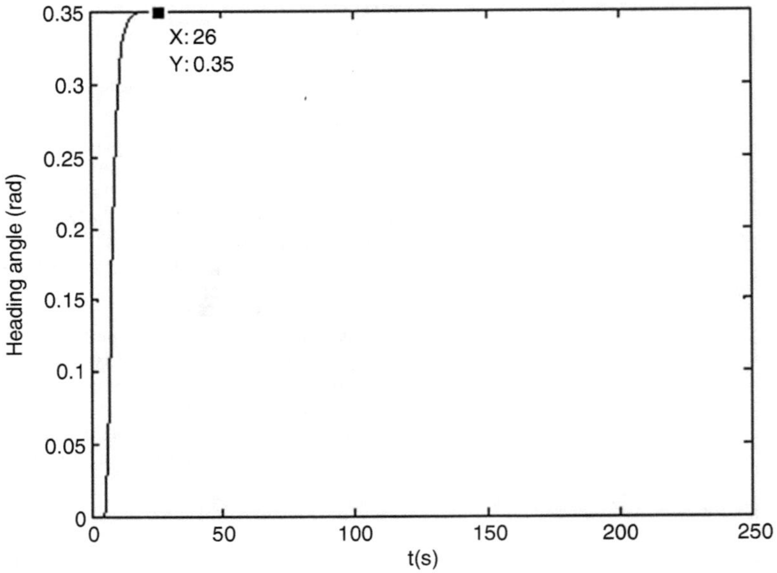

Fig. 9.27 Response of the system to a 0.35 radians reference heading input

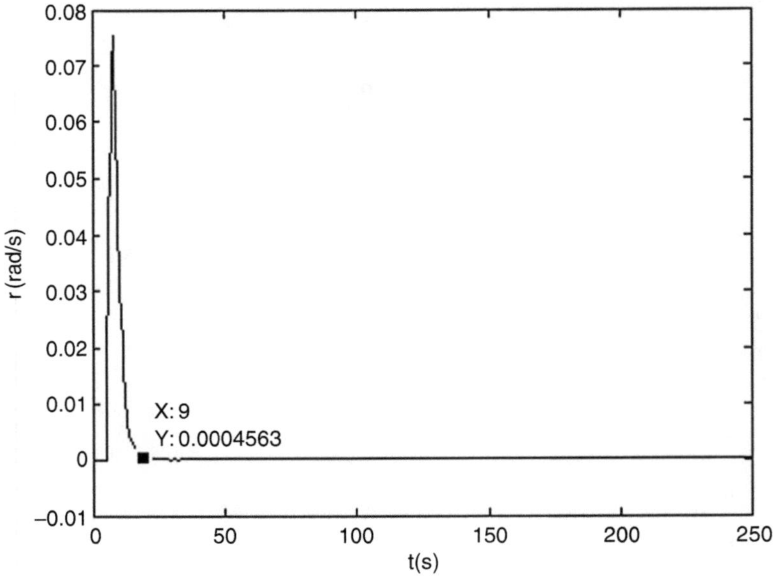

Fig. 9.28 Change in r for a heading input of 0.35 radians

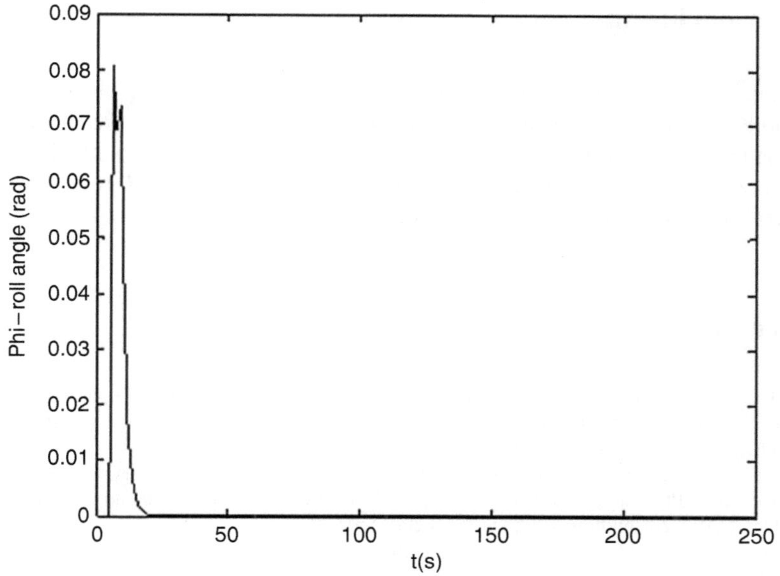

Fig. 9.29 Change in Φ for a heading input of 0.35 radians

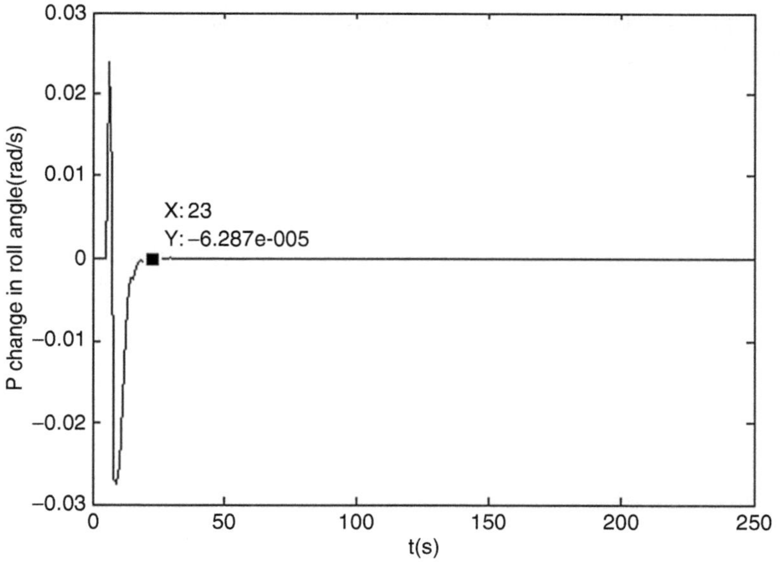

Fig. 9.30 Change in p (rad/s) for a heading input of 0.35 radians

9.4 Conclusion

In this chapter, a flight control system is designed for UAVs using classical control methods. Separate discussions for controlling different parameters, such as altitude and heading, are given, and controller design procedures are explained in detail. This method is the most common method for UAV control, and good results for controlling both the lateral and longitudinal motion parameters also prove that.

References

1. Golnaraghi F, Kuo BC (2009) Automatic control systems, 9th edn. Wiley, Englewood Cliffs
2. Yechout TR, Morris SL, Bossert DE, Hallgren WF (2003) Introduction to aircraft flight mechanics. AIAA education series, Reston
3. Williams W (2003) UAV handling qualities.....you must be joking. Aerospace Sciences Corporation Pty. Ltd., Elizabeth, Australia. Available online at: http://aerosciences.com.au/hidden/UAV%20Handling%20Qualities%20Paper%20v1.pdf
4. Capello E, Guglieri G, Marguerettaz P, Quagliotti F (2012) Preliminary assessment of flying and handling qualities for mini-UAVs. J Intell Robot Syst 65(1–4):43–61. doi:10.1007/s10846-011-9565-5
5. Pamadi BN (2003) Performance, stability, dynamics, and control of airplanes. AIAA education series, Reston
6. Nelson RC (1998) Flight stability and automatic control. McGraw-Hill, New York
7. Haugen F (2004) PID control. Tapir Academic Press, Trondheim
8. Stevens BL, Lewis FL (1992) Aircraft control and simulation. Wiley, Chichester

Chapter 10
Linear Quadratic Regulator Controller Design

10.1 Introduction

The first steps to design a control system for an UAV is defining the dynamic model and finding the aerodynamic coefficients. The nonlinear model can then be linearized to design a linear controller for the UAV [1]. The characteristic values of different motions that show how the aircraft normally behaves must also be investigated at this stage. Finally, the controller and filter can be designed considering the required specifications.

In this chapter, the optimal control technique linear quadratic regulator (LQR) is chosen to design a controller for the longitudinal motion of a small fixed-wing type UAV. The Kalman filter technique is then applied to see how the controller is affected by disturbances. The effectiveness of the controller with and without the Kalman filter is also shown through simulations. Linear quadratic control is an optimal control technique that is used for controlling the aircraft. This control technique aims to decrease the energy that is used to control the aircraft. This technique can be applied together with a filtering technique (Kalman filter, etc.) in cases where some of the states are not available for measurement or when the measurements are noisy.

Studies that include LQR control design for different types of aircraft have been carried out in the literature [2]. In [3], an LQR method is used to control the trajectory and mission paths of the autonomous helicopter. Nonlinear motion dynamics are linearized at certain operating points and a linear model is obtained by Taylor series expansion. By using the LQR methodology, the attitude of the autonomous Puma helicopter is controlled. The study by Öner et al. [4] shows that the LQR controller is quite effective in the vertical flight mode for all possible yaw angles. In [5], the design procedure for a gain-scheduled LQR controller for an autonomous airship is presented. Two types of control subsystems (lateral and longitudinal) have been designed from the nonlinear six degrees of freedom (6DOF)

C. Hajiyev et al., *State Estimation and Control for Low-cost Unmanned Aerial Vehicles*, DOI 10.1007/978-3-319-16417-5_10

airship model to fulfill different goals (yaw as well as speed and position control). In references [2–5], the LQR controller is not applied together with a filtering technique.

In some studies, filtering techniques are used to analyze the effects of LQR design and increase the effectiveness of the controller. The paper by Santoso et al. [6] depicts the application of linear quadratic optimal control to the longitudinal flight motion of a UAV which has elevon control only. The LQR controller is followed by a Kalman filter-based estimator for unmeasurable states. The LQR controller is then combined with the Kalman estimator using the separation principle to investigate the feasibility of altitude control. The simulation results show improvements compared with classical design counterparts in the sense that the combined approach offers more design flexibility and is able to tolerate noisy environments.

Some advanced techniques, including robust and adaptive control techniques, are also used in UAV control. For example, in one study by Johnson et al., the H-infinity technique is used to design a velocity and altitude controller that follows a determined model [7]. However, in this study, these advanced techniques are only investigated to design a better controller and are not used.

In this chapter, an altitude controller is designed for a small UAV using the optimal LQR control method with the Kalman estimator. The effectiveness of the controller with the gains found by the LQR method is investigated under the effect of disturbances and with the Kalman filter.

10.2 Linear Quadratic Optimal Controller

In order to design an optimal controller for the UAV, we use the linear quadratic controller method in this chapter. This method depends on minimizing criteria using the feedback $u = -kx(t)$. The performance index that is used for minimizing both the control effort and the states can be given as [8]:

$$J = \frac{1}{2}\int_{0}^{\infty} \left(x^{\mathrm{T}}Qx + u^{\mathrm{T}}Ru\right)\mathrm{d}t \tag{10.1}$$

$\|x\| = x^{\mathrm{T}}(t)x(t)$ and x is an $n \times 1$ state vector. The given norm that depends on the states can be used as a measure of the system response. The index also includes u inputs to control the system. The Q and R matrices in J are called the weight matrices. The weight matrices allow us to define the effect of each state and control input in the controller response when we use the feedback for designing the controller.

Q is an $n \times n$ real, symmetric, positive (or semipositive) definite matrix. Q can be given as:

$$Q = \begin{bmatrix} q_1 & 0 & \cdots & 0 \\ 0 & q_2 & & 0 \\ \vdots & & \ddots & 0 \\ 0 & 0 & 0 & q_n \end{bmatrix} \tag{10.2}$$

The weight given to the ith element of the matrix Q is a measure of the control effort to control the related state. The bigger the element, the bigger the control effort used for that state.

The matrix R is a $p \times p$, real, symmetric, and positive-definite matrix. Choosing a positive-definite R is a must for an optimal solution.

If we can find the feedback $u = -kx(t)$ that minimizes the J index for the system given by $\dot{x} = Ax + Bu$, then we can solve the optimal control problem. In order to do this, we need to relate the Lyapunov stability criteria with the quadratic performance index.

10.2.1 Lyapunov Stability Criteria

The Lyapunov stability criteria deal with the behavior of the system response. When the system leaves a steady state, whether it returns to a stable point or not can be measured using the Lyapunov criteria. If a nonvibratory system response takes the system very close to a point where it is stable for every x, then the system is called Lyapunov stable. If the nonvibratory response of a system from a point x goes near point x^e (equilibrium point) for every x, then the system is called Lyapunov stable. If the system response goes to x^e for every x, then the system is called Lyapunov asymptotically stable.

In order to control this criterion, the norm values $\|x\| = x^{\mathrm{T}}(t)x(t)$ that produce a hypercubic space near the equilibrium point can be used. If the system response stays in an area defined as S, it shows that the system is stable. In conclusion, using the norm value, the total energy in the system can be calculated and the stability condition can be checked. This method is called the Lyapunov method.

Generally, Lyapunov-type scalar functions can be defined as:

$$V(x) = x^{\mathrm{T}}Px \tag{10.3}$$

where P is a real, symmetric, positive- or negative-definite matrix. The positive definiteness of the matrix P defines the status of the Lyapunov function. If a

Lyapunov function can be built for the system and the function has partial derivatives for every x variable, the stability condition can be checked using these partial derivatives.

The positive definiteness requirements are:

$$V(x) > 0 \tag{10.4}$$

$$V(0) = 0 \tag{10.5}$$

The following conditions are checked in order to understand whether the system is stable or not:

(a) Asymptotically stable if $\dot{V}(x) < 0$ and $x \neq 0$
(b) Generally asymptotically stable if the requirements in (a) are met and $V(x) -> \infty$ when $\|x\| -> \infty$

Thus, a quadratic Lyapunov function can be defined and used for checking system stability. This leads us to the linear quadratic optimal control method.

10.2.2 Linear Quadratic Optimal Control Using Lyapunov Criterion

When we add $u = -kx(t)$, the performance index J becomes:

$$J = \frac{1}{2}\int_0^\infty \left(x^T Q x + x^T K^T R K x\right) dt \tag{10.6}$$

A Lyapunov stability function $V(x(t))$ can be defined similar to the J performance index:

$$V(x(t)) = \frac{1}{2}\int_0^\infty \left(x^T Q x + x^T K^T R K x\right) dt \tag{10.7}$$

The derivative of the Lyapunov function can be found, keeping in mind that $x(\infty) = 0$:

$$\dot{V}(x(t)) = -\frac{1}{2}\left(x^T Q x + x^T K^T R K x\right) \tag{10.8}$$

The Lyapunov function can also be written as:

$$V(x) = \frac{1}{2}x^T P x \tag{10.9}$$

The derivative of this function is:

$$\dot{V}(x) = \frac{1}{2}\left(x^{\mathrm{T}} P x + x^{\mathrm{T}} P \dot{x}\right) \tag{10.10}$$

Combining the equations found for the derivative of $V(x)$, the following equation can be solved:

$$(A - BK)^{\mathrm{T}} P + P(A - BK) + K^{\mathrm{T}} R K + Q = 0 \tag{10.11}$$

The value of $V(x)$ at 0 is equal to the performance index value for the system trajectory starting from 0. Thus, the J index can be calculated as:

$$J = \frac{1}{2} x^{\mathrm{T}}(0) P x(0) \tag{10.12}$$

The values of matrix P that minimize the performance index J can then be calculated as:

$$\frac{\partial\left(x^{\mathrm{T}}(0) P x(0)\right)}{\partial k_{ij}} = 0 \tag{10.13}$$

If the values of P that are optimal make the matrix P positive-definite, this guarantees a stable loop defined by the Lyapunov function and we can calculate the optimal gain. We may obtain the same result using the Riccati equation, which is:

$$R = \Gamma^{\mathrm{T}} \Gamma, \tag{10.14}$$

from the definition of the matrix R. Using the given equation for R and deriving P to find the optimal values, we obtain the optimal control input:

$$\frac{\partial}{\partial k_{ij}}\left(\Gamma K - \left(\Gamma^{\mathrm{T}}\right)^{-1} B^{\mathrm{T}} P\right)^{\mathrm{T}}\left(\Gamma K - \left(\Gamma^{\mathrm{T}}\right)^{-1} B^{\mathrm{T}} P\right) = 0 \tag{10.15}$$

This equation is calculated using the Lyapunov-type formula. Deriving for K, we can find u, the input that brings the system optimally under control:

$$\Gamma K = \left(\Gamma^{\mathrm{T}}\right)^{-1} B^{\mathrm{T}} P \tag{10.16}$$

$$K = R^{-1} B^{\mathrm{T}} P \tag{10.17}$$

$$u(t) = -K R^{-1} B^{\mathrm{T}} P x(t) \tag{10.18}$$

The Riccati matrix equation can be found by using the previously determined value of K [8].

$$A^{\mathrm{T}}P + PA - PBR^{-1}B^{\mathrm{T}}P + Q = 0 \tag{10.19}$$

For designing a controller, both the first formula and the given Riccati equation can be used. If the Riccati equation is to be used, the longitudinal matrices A_{lon}, B_{lon} and lateral matrices A_{lat}, B_{lat} must first be determined. Then, the equation for finding the matrix P can be determined or MATLAB functions can be used to find the required values.

As an advantage, the LQR design can be done easily. However, a disadvantage of this method is expecting that the values of the states are known exactly at every moment. However, in general, these values are not known exactly and a filtering technique is necessary. In this study, Kalman filtering is used to estimate the states that are needed in the optimal controller.

10.3 Altitude and Speed Controller Design Using the LQR Method

10.3.1 LQR Altitude Controller

Using the LQR method, a longitudinal controller for altitude and speed and a lateral controller for heading is designed. The structure of the system allows us to use the gain values found using the Riccati equation directly. The formulas given below describe the motion of the vehicle around the stable point. Here, the $\tilde{x}$ represents the divergence from the required point and xd represents the reference input:

$$\tilde{x} = x - xd \tag{10.20}$$

$$\dot{\tilde{x}} = \dot{x} - \dot{xd} \tag{10.21}$$

$$\dot{\tilde{x}} = Ax + Bu - \dot{xd} \tag{10.22}$$

$$\dot{\tilde{x}} = Ax + Bu - \dot{xd} + Axd - Axd \tag{10.23}$$

$$\dot{\tilde{x}} = A\tilde{x} + Bu + Axd - \dot{xd} \tag{10.24}$$

As we can understand, if $Axd = \dot{xd}$ holds using only $u = -kx(t)$ without adding anything into the state matrices, it will be enough to control the system optimally. With choosing $xd = [0; 0; 0; 0; hd]$, where hd is the reference altitude input, we can reach the reference point only using state feedback in the given system. In short, in cases where we have to control the state which has no direct effect on the other states (where the A matrix columns related to the state are zeros), we can use the given method directly.

The scheme of the controller is given in Fig. 10.1. In the scheme, *hd* values are multiplied by $K_1 = [0;0;0;0;1]$ and added to $u = -k_{LQR}x(t)$ to find the real control input that is used in the state model. The longitudinal states h, u, w, q, and θ are scoped.

The controller is used to take the altitude value to the reference while holding the other values constant. As we can see in Figs. 10.2, 10.3, 10.4, 10.5, 10.6, and 10.7, the changes in the values other than the altitude goes to zero and the states hold their values.

The weight matrices (Q, Q_1, and R) used in the controller are given below. The change in U observed while the altitude value is changing can be controlled by changing the corresponding weight matrix value. For this purpose, two different Q matrices (Q and Q_1) are used and the results are given. In R, the value for the elevator is chosen to be high in order to use the elevator effort more effectively:

$$R = \begin{bmatrix} 100 & 0 \\ 0 & 1 \end{bmatrix} \tag{10.25}$$

$$Q = \begin{bmatrix} 1 & 0 & 0 & 0 & 0 \\ 0 & 1 & 0 & 0 & 0 \\ 0 & 0 & 1 & 0 & 0 \\ 0 & 0 & 0 & 100000 & 0 \\ 0 & 0 & 0 & 0 & 10 \end{bmatrix} \tag{10.26}$$

$$Q_1 = \begin{bmatrix} 10 & 0 & 0 & 0 & 0 \\ 0 & 1 & 0 & 0 & 0 \\ 0 & 0 & 1 & 0 & 0 \\ 0 & 0 & 0 & 100000 & 0 \\ 0 & 0 & 0 & 0 & 10 \end{bmatrix} \tag{10.27}$$

10.3.2 LQR Speed Controller

To design an LQR-type speed controller, an integrator is added to the system. For this system, which includes forward velocity and integrator states, a gain value using the given LQR method is obtained. In Fig. 10.8, the system built in Simulink for simulation purposes can be seen. Two controllers are needed for the longitudinal control in order to control both speed and altitude. A MUX function is used to add the integrator state to the system and the u input values are fed through actuator functions and the integrator. After adding the integrator, the gain value K_{lon} is multiplied by system states and supplied to the system model as inputs.

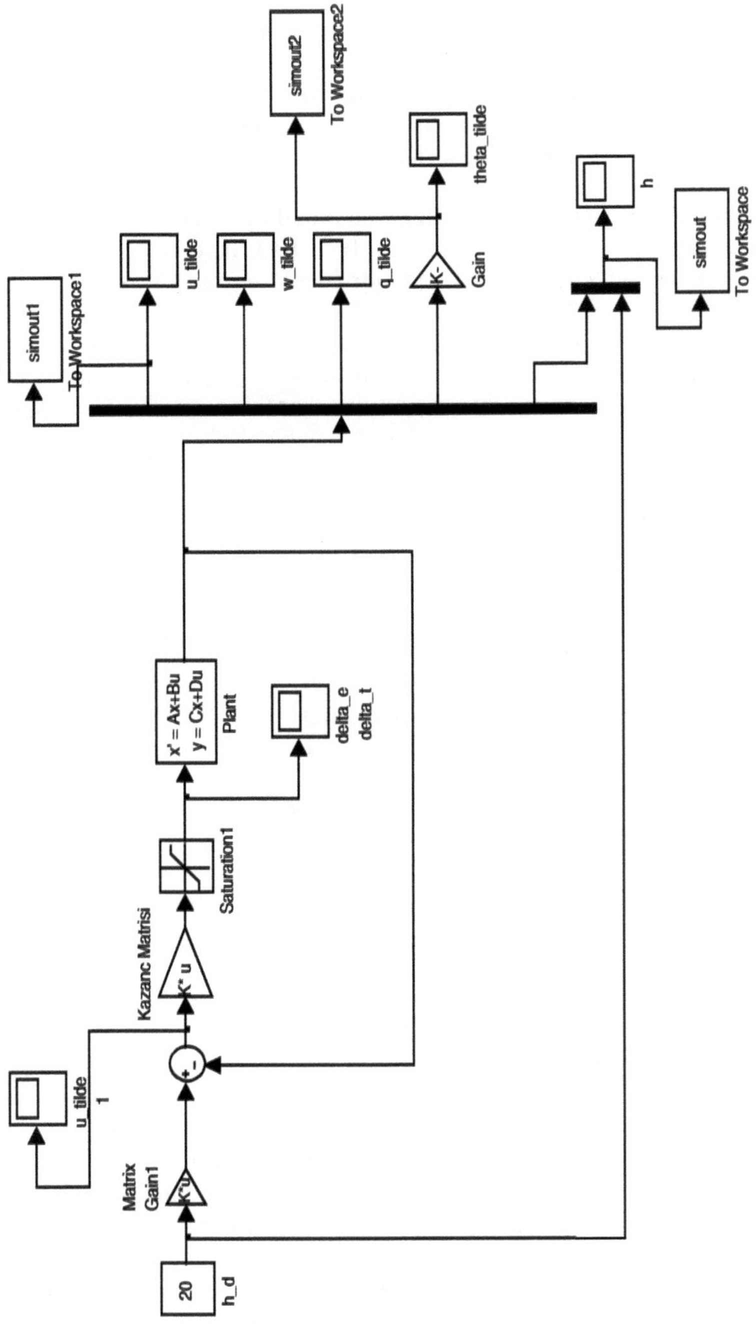

Fig. 10.1 Linear quadratic regulator (LQR) altitude controller

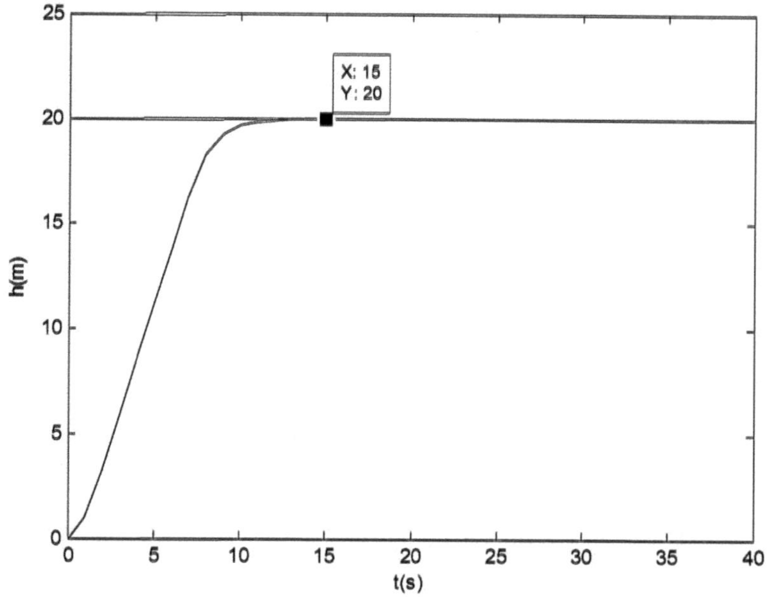

Fig. 10.2 LQR altitude controller response for a 20-m *hd* input (with *Q*)

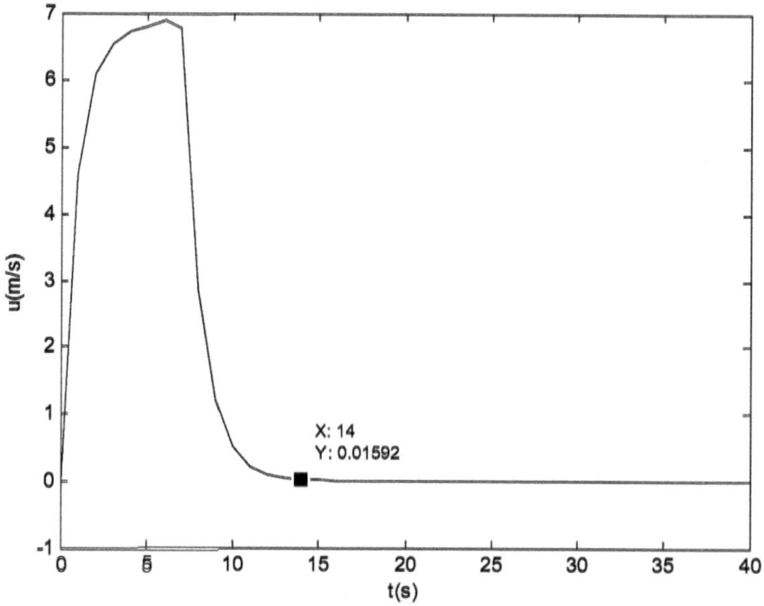

Fig. 10.3 Change in *U* for a 20-m altitude input (with *Q*)

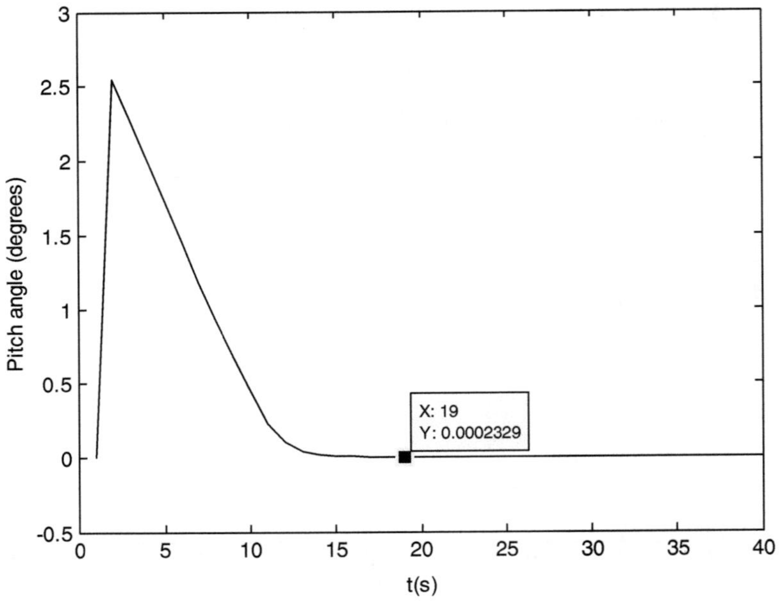

Fig. 10.4 Change in pitch angle for a 20-m *hd* input (with Q)

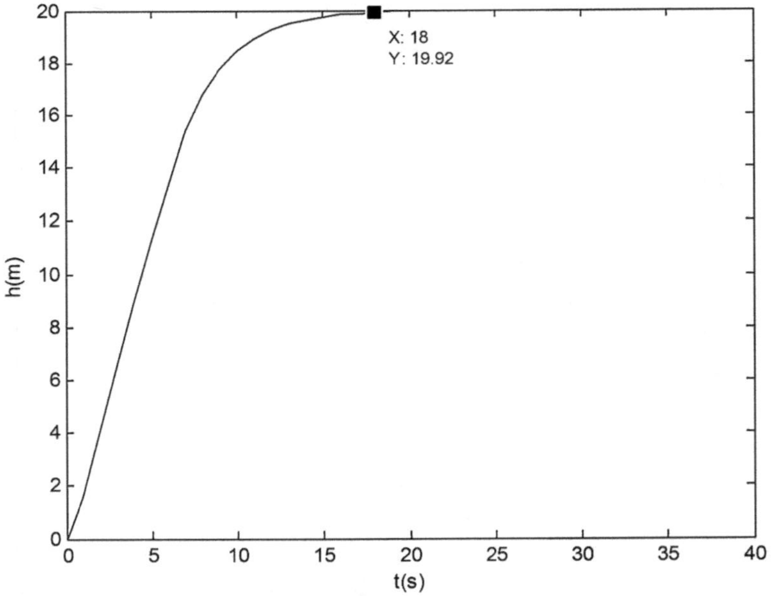

Fig. 10.5 LQR altitude controller response for a 20-m *hd* input (with Q_1)

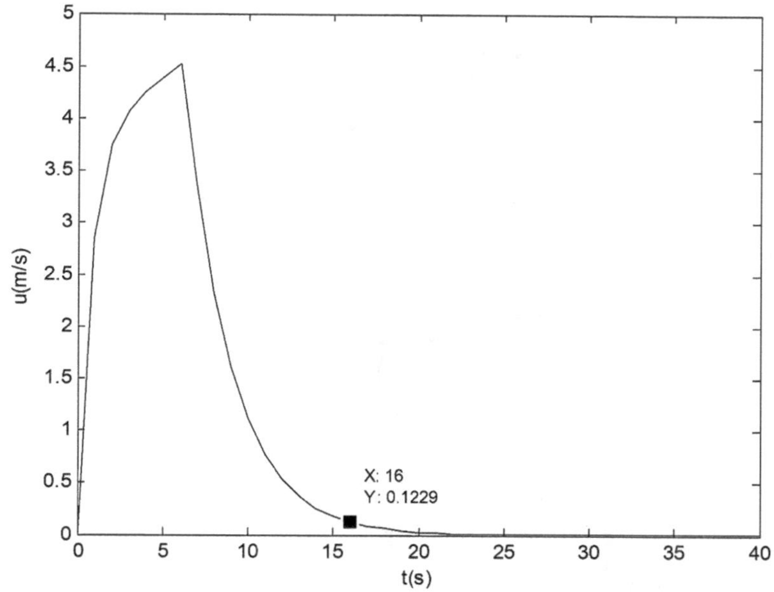

Fig. 10.6 Change in U for a 20-m altitude input (with Q_1)

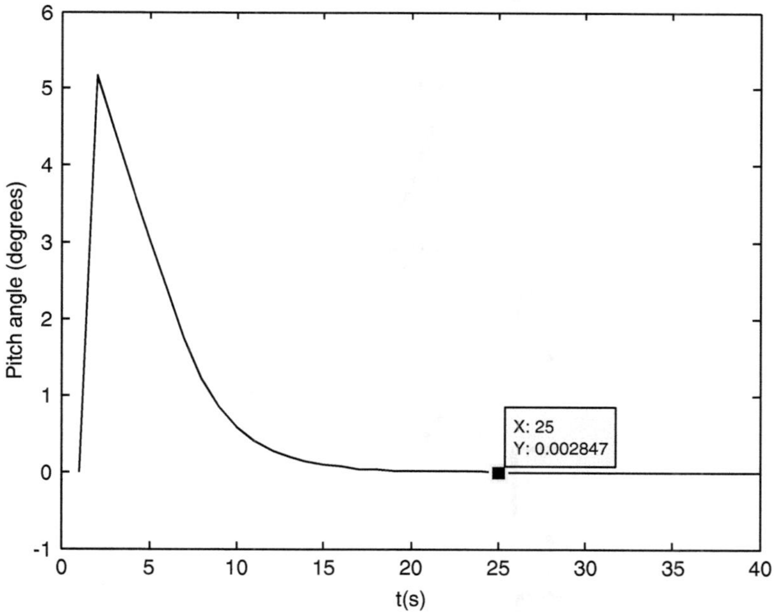

Fig. 10.7 Change in pitch angle for a 20-m *hd* input (with Q_1)

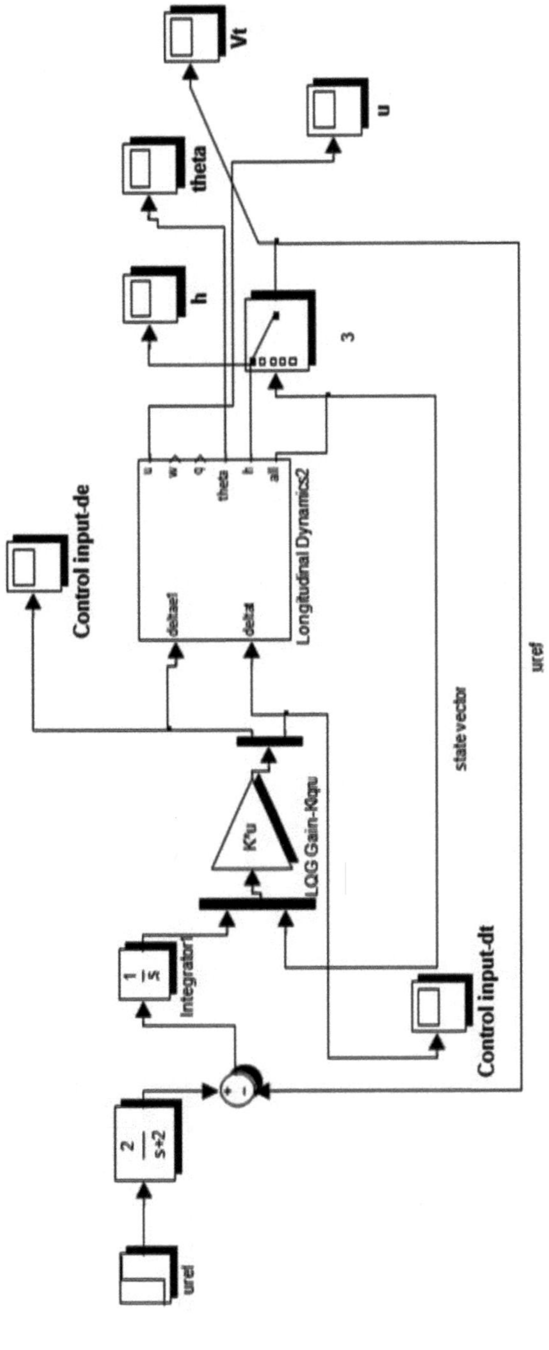

Fig. 10.8 Simulink scheme of the forward velocity controller

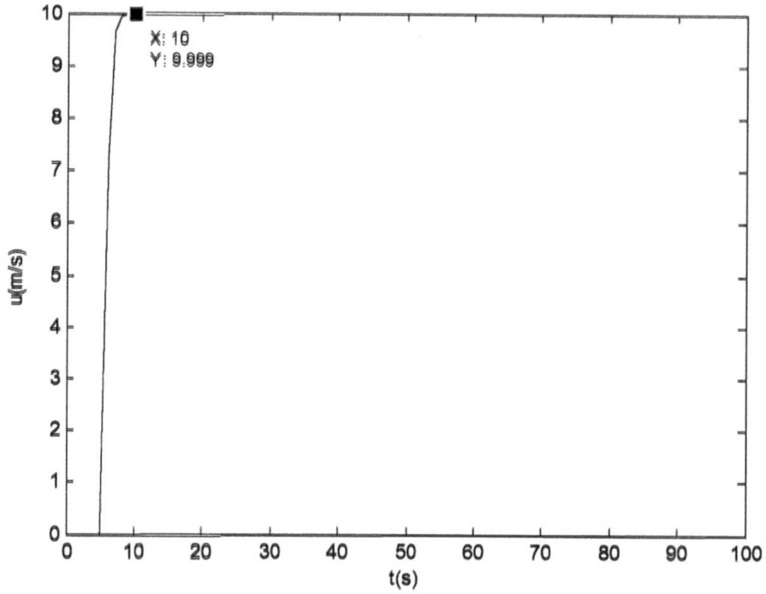

Fig. 10.9 Response of the LQR speed controller to a 10-m/s reference input

The designed speed controller takes the aircraft to the reference speed while holding the other states constant. The changes in different states when a 10-m/s change in U is commanded are given in Figs. 10.9, 10.10, and 10.11.

The integrator is added to the system as the first state. The error of the speed is fed to the system as a sixth state in the longitudinal equations. The new A and B matrices built by adding the integrator are given below:

$$A_{\text{lon}} = \begin{bmatrix} 0 & -1.0000 & 0 & 0 & 0 & 0 \\ 0 & -0.3356 & 1.3181 & -1.9276 & -9.6610 & 0 \\ 0 & -1.7916 & -3.9003 & 9.8215 & -1.7035 & 0 \\ 0 & 0.7020 & -3.5375 & -11.3920 & 0 & 0 \\ 0 & 0 & 0 & 1.0000 & 0 & 0 \\ 0 & -0.1736 & -0.9848 & 0 & 17.4865 & 0 \end{bmatrix} \tag{10.28}$$

$$B_{\text{uzu}} = \begin{bmatrix} 0 & 0 \\ -0.7436 & 6.8728 \\ 3.7855 & 0 \\ 47.9170 & 0 \\ 0 & 0 \\ 0 & 0 \end{bmatrix} \tag{10.29}$$

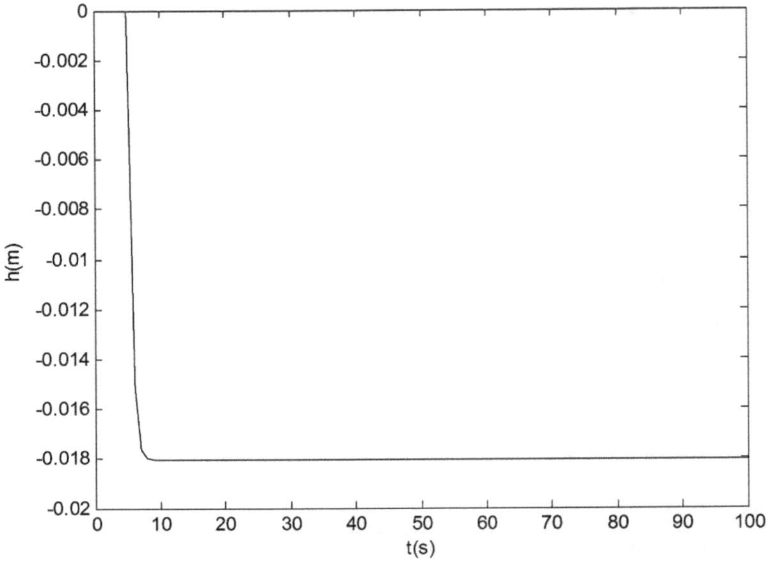

Fig. 10.10 Change in altitude when the LQR speed controller takes the system to 10 m/s

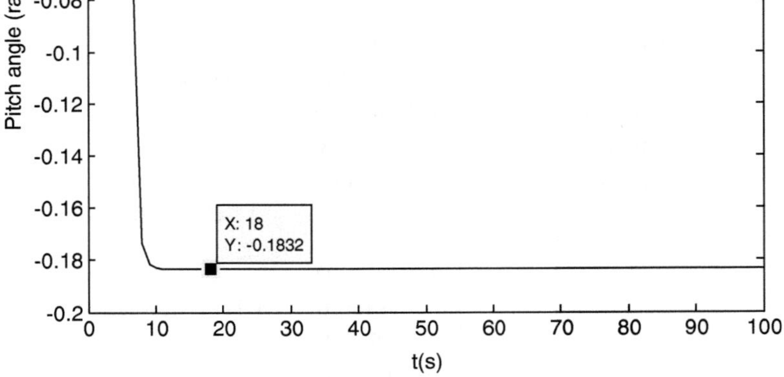

Fig. 10.11 Change in pitch angle when the LQR speed controller takes the system to 10 m/s

It is observed in the tryouts that the reference speed value can be reached quickly without affecting the other states (altitude, pitch angle, etc.) when we choose the diagonal elements of Q_{lon} as 1000, 10, 1, 1, 100, and 30000. The diagonal elements of R_{lon} are chosen as 100 and 100 to give the same effort to both elevator and thrust control inputs:

$$Q_{\mathrm{uzu}} = \begin{bmatrix} 1000 & 0 & 0 & 0 & 0 & 0 \\ 0 & 10 & 0 & 0 & 0 & 0 \\ 0 & 0 & 1 & 0 & 0 & 0 \\ 0 & 0 & 0 & 1 & 0 & 0 \\ 0 & 0 & 0 & 0 & 100 & 0 \\ 0 & 0 & 0 & 0 & 0 & 30000 \end{bmatrix} \tag{10.30}$$

$$R_{\mathrm{uzu}} = \begin{bmatrix} 100 & 0 \\ 0 & 100 \end{bmatrix} \tag{10.31}$$

10.4 LQR-Type Heading Controller

For the lateral controller, an integrator may also be added to the system; however, the structure of the system tells us that, for controlling the heading using the reference input $xd = [0; 0; 0; 0; \psi d]$, the normal lateral matrices are sufficient. For the case where heading is the input, the following equation holds:

$$\dot{xd} = A_{\mathrm{lat}} x_d \tag{10.32}$$

Thus, as stated, we can use the lateral state and control distribution matrices to calculate the result of the Riccati equation and the LQR gain values. The Q and R matrices for the lateral LQR controller are chosen as follows:

$$Q = \begin{bmatrix} 1 & 0 & 0 & 0 & 0 \\ 0 & 10 & 0 & 0 & 0 \\ 0 & 0 & 100 & 0 & 0 \\ 0 & 0 & 0 & 10000 & 0 \\ 0 & 0 & 0 & 0 & 1000 \end{bmatrix} \tag{10.33}$$

$$R = \begin{bmatrix} 100 & 0 \\ 0 & 100 \end{bmatrix} \tag{10.34}$$

The scheme of the lateral LQR heading controller is given in Fig. 10.12. As we can understand from the results (Figs. 10.13, 10.14, and 10.15), the LQR controller is working optimally and takes the UAV to the desired heading with little changes occurring in δ_a and δ_r. The vibrations in states may be lessened by choosing appropriate Q and R weight matrices.

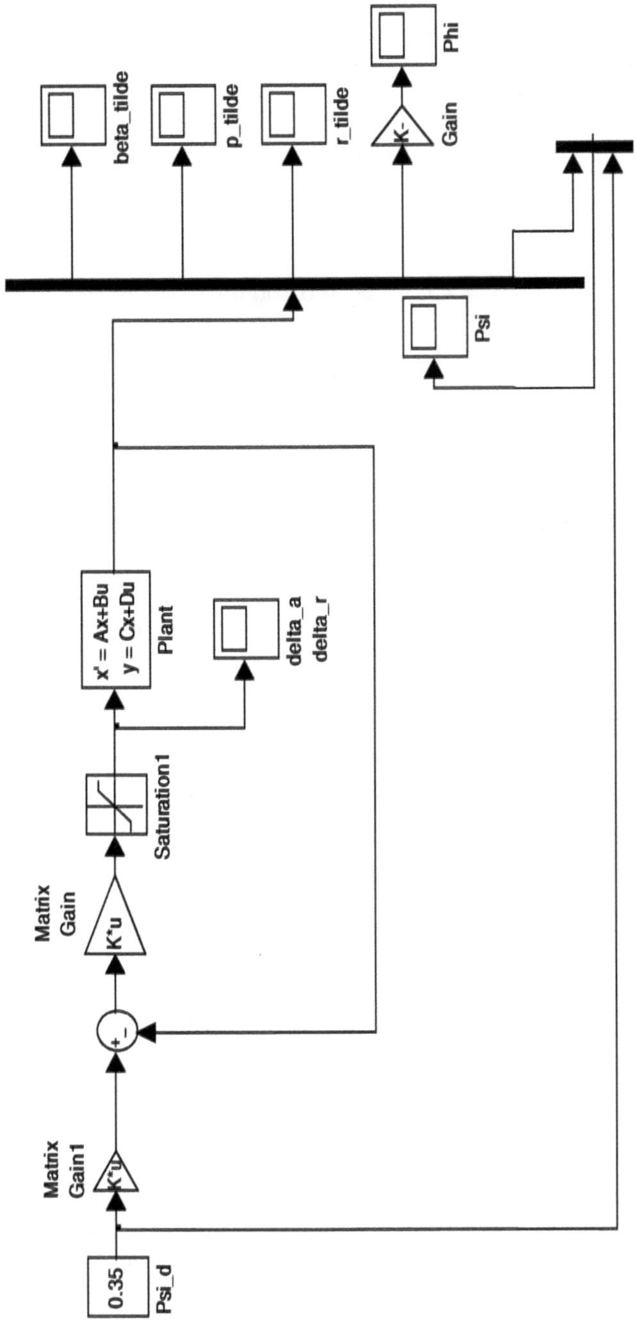

Fig. 10.12 Lateral LQR controller: MATLAB Simulink scheme

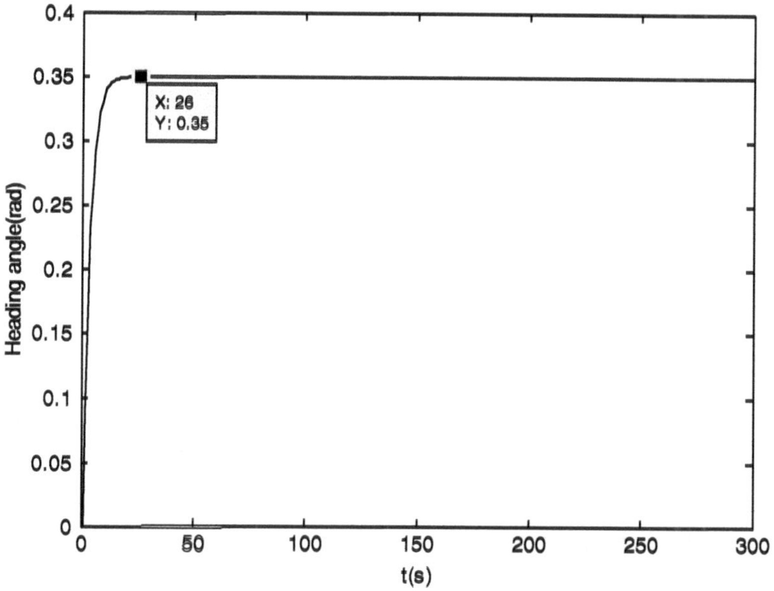

Fig. 10.13 LQR controller response for a $\psi d = 0.35$ radians input

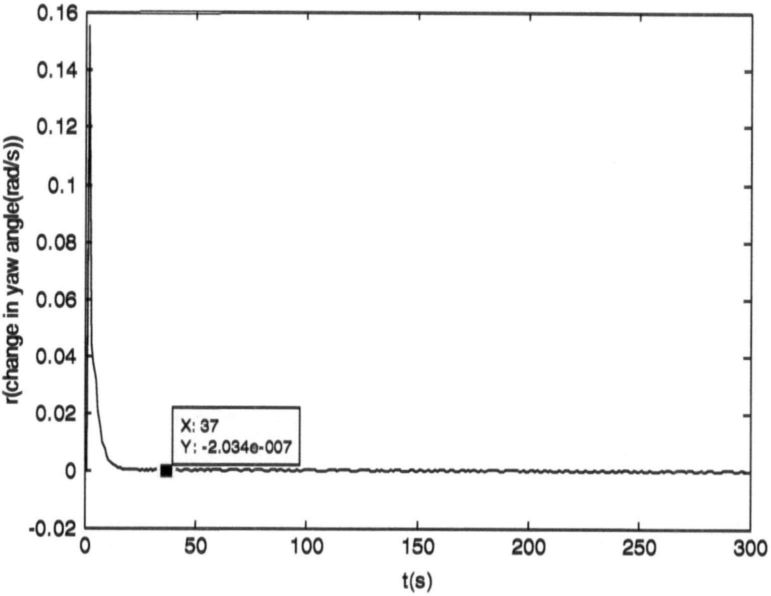

Fig. 10.14 Change in r (yaw angle change, rad/s) when a $\psi d = 0.35$ radians input is given

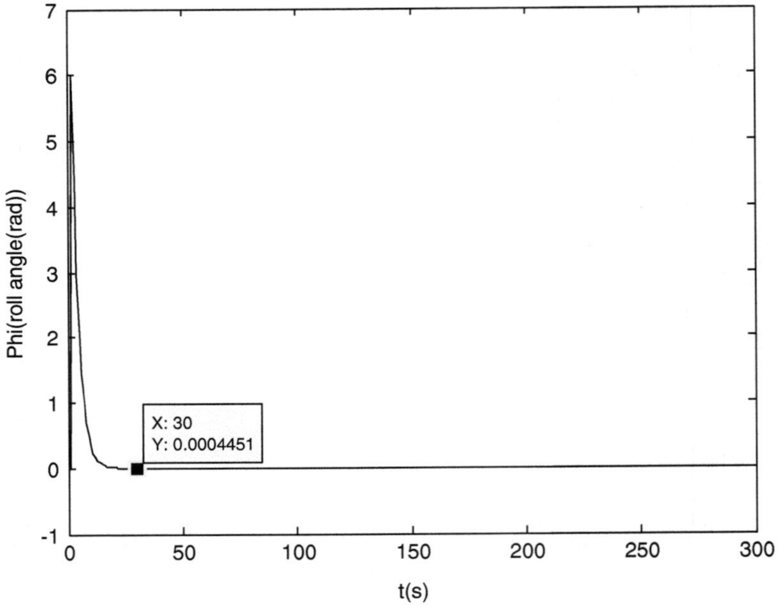

Fig. 10.15 Change in roll angle (Φ) when a $\psi d = 0.35$ radians input is given

10.5 LQR Controller with the Kalman Estimator

In this section, it is assumed that the measured states are correct at any time and the designed LQR is fed with them to control the UAV. However, noise is always present and the system may not give accurate results under normal operating conditions. For this purpose, a Kalman filter that has the capability to estimate the real values of the states is incorporated in the study. The results of the system with a Kalman filter and LQR and only LQR are then compared to show the effectiveness of the Kalman filter. The system model is discretized using the Euler method. Firstly, the nominal model is tested with 20 m altitude and 20.02° heading inputs, with the optimal gain values calculated previously. Then, the system with noise is simulated. The response of the system with the controller under the disturbances is tested with and without the Kalman filter and the effectiveness of the designed LQR and Kalman filter system is shown [9].

The Kalman filter uses state equations (state space matrices) and initial values to calculate the residue and gain values and to estimate the real signal value. The steps of the Kalman filter can be explained using the linear discretized state and measurement equations:

$$X(k+1) = AX(k) + Bu(k) + Gw(k) \tag{10.35}$$

$$y(k) = HX(k) + v(k) \tag{10.36}$$

In the state equation, $X(k)$ is the state vector of the system, A is the system transition matrix, $u(k)$ is the input vector, B is the control distribution matrix, $w(k)$ is the random Gaussian noise vector (system noise) with zero mean and known covariance structure, and G is the transition matrix of the system noise. In the measurement equation, $y(k)$ is the measurement vector, H is the measurement matrix, and $v(k)$ is the measurement noise vector with zero mean and known covariance structure.

There is no correlation between the system noise $w(k)$ and the measurement noise $v(k)$. The covariance matrices for the $w(k)$ and $v(k)$ vectors are given by:

$$E\left[w(k)w^{\mathrm{T}}(j)\right] = Q(k)\delta(kj)$$

$$E\left[v(k)v^{\mathrm{T}}(j)\right] = R(k)\delta(kj)$$

Here, E is the expected value operator and $\delta(kj)$ is the Kronecker symbol.

The optimum linear Kalman filter that estimates the state vector of the system (10.35) is expressed using the following recursive equations system:

Equation of the extrapolation value:

$$X_e(k/k-1) = AX_e(k-1/k-1) + BK_{\mathrm{LQR}}(k-1)(Xd - X_e(k-1/k-1)) \tag{10.37}$$

The innovation sequence:

$$\Delta(k) = Z(k) - HX_e(k/k-1) \tag{10.38}$$

Equation of the estimation value:

$$X_e(k/k) = X_e(k/k-1) + K(k)\Delta(k) \tag{10.39}$$

Gain matrix of the optimum linear Kalman filter:

$$K(k) = P(k/k)H^{\mathrm{T}}R^{-1}(k) = P(k/k-1)H^{T}\left(HP(k/k-1)H^{T} + R(k)\right)^{-1} \tag{10.40}$$

The covariance matrix of the filtering error is:

$$P(k/k) = (I - K(k)H)P(k/k-1) \tag{10.41}$$

The covariance matrix of the extrapolation error is:

$$P(k/k-1) = AP(k-1/k-1)A^{\mathrm{T}} + BD_u(k-1)B^{\mathrm{T}} + GQ(k-1)G^{\mathrm{T}} \tag{10.42}$$

where Xd is the desired vector and I is the identity matrix.

The Kalman filter tries to estimate the real signal from the signal with disturbance, which has a Gaussian distribution, using the described steps and decreasing the value between two signals [10].

10.5.1 Longitudinal LQR with the Kalman Estimator

The model of longitudinal motion can be discretized using the Euler approach for a sampling time of dt. Thus, the new discretized A and B matrices to be used in the filtering approach can be found by $A_1 = A \times dt + I$ and $B_1 = B \times dt$.

The discretized UAV model can be given as follows:

$$X(k+1) = \begin{bmatrix} 0.997 & 0.013 & -0.02 & -0.097 & 0 \\ -0.018 & 0.961 & 0.098 & -0.017 & 0 \\ 0.007 & -0.035 & 0.886 & 0 & 0 \\ 0 & 0 & 0.01 & 1 & 0 \\ -0.0017 & -0.01 & 0 & 0.175 & 1 \end{bmatrix} \begin{pmatrix} u(k) \\ w(k) \\ q(k) \\ \theta(k) \\ h(k) \end{pmatrix} + \begin{bmatrix} -0.007 & 0.07 \\ 0.04 & 0 \\ 0.48 & 0 \\ 0 & 0 \\ 0 & 0 \end{bmatrix} \begin{bmatrix} \delta_e(k) \\ \delta_t(k) \end{bmatrix} + Gw(k) \quad (10.43)$$

$$y(k) = HX(k) + v(k) \quad (10.44)$$

In our case, disturbance with Gaussian white noise characteristics generated by MATLAB commands is applied to the real values found using the model of the longitudinal dynamics of the UAV. The Kalman filtering technique is then applied and its effectiveness is shown. In the real scenario, disturbances in measurements and processes are normal and have an effect on the controller. Thus, using a filtering technique is very important.

The values of the states can be calculated using Eq. (10.43), which includes the control rule. The gain value K is the one found for the longitudinal LQR controller case. The disturbances on the states must, of course, be first determined and applied to the system. Finally, the Kalman filter can be applied to the system with the disturbances to develop an effective controller. To do this, some MATLAB code is written. The results are given in Figs. 10.16, 10.17, 10.18, 10.19, 10.20, 10.21, 10.22, 10.23, and 10.24. The Kalman filter that works as an optimal observer estimates the new values of the states correctly and decreases the error [10, 11].

As can be seen from Figs. 10.16, 10.17, 10.18, 10.19, 10.20, 10.21, 10.22, 10.23, and 10.24, in case of disturbances, using a Kalman filter to estimate the values of the longitudinal states clearly increases the effectiveness of the LQR controller.

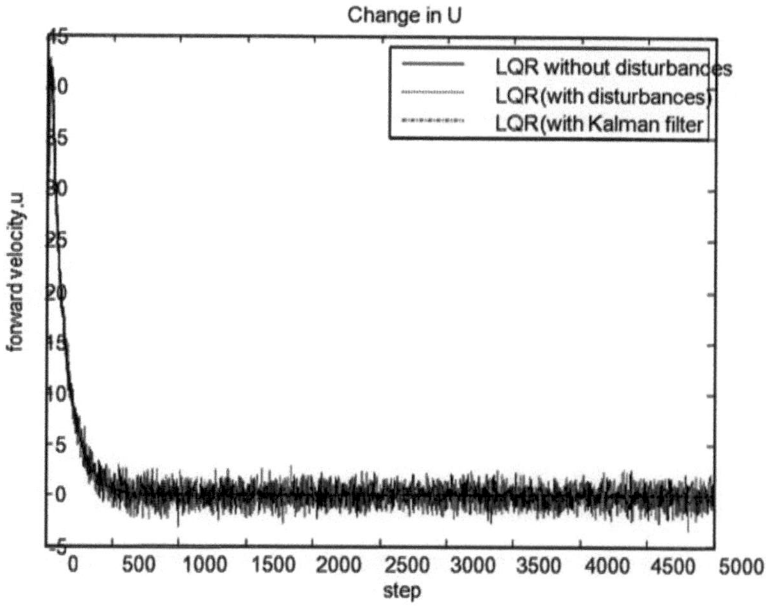

Fig. 10.16 Change in U velocity (5,000 steps = 500 s)

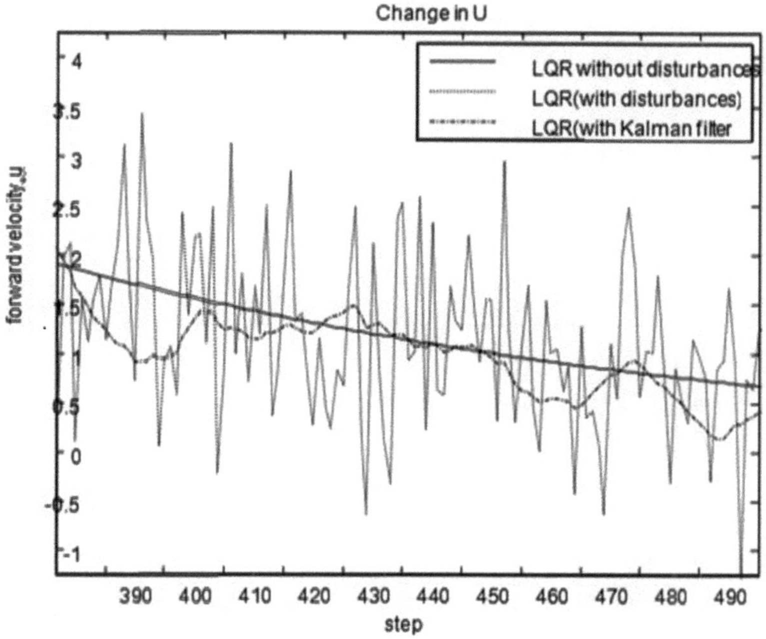

Fig. 10.17 Change in U velocity (zoomed, 5,000 steps = 500 s)

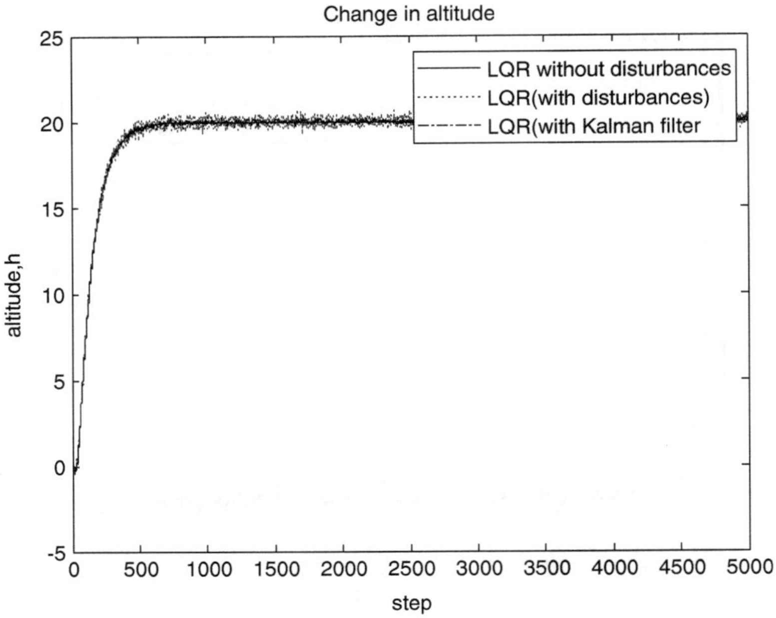

Fig. 10.18 Change in h in the LQR controller (5,000 steps = 500 s)

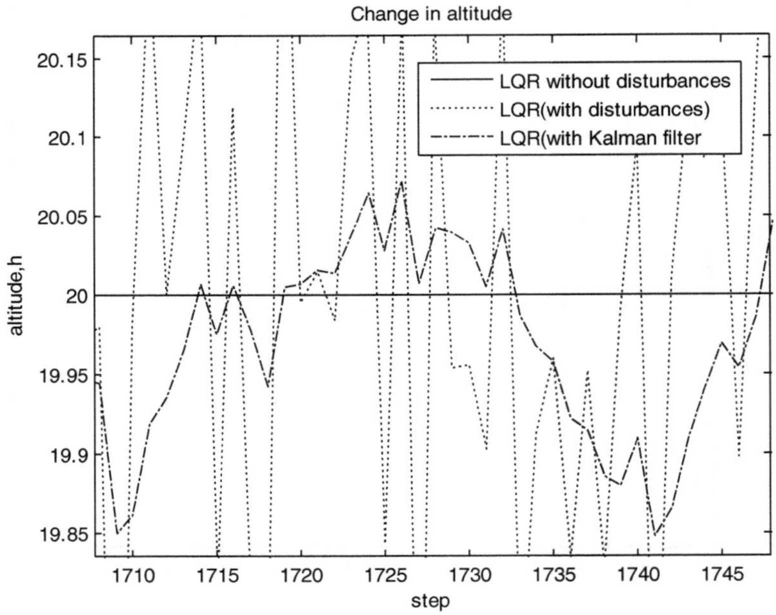

Fig. 10.19 Change in h in the LQR controller (zoomed, 5,000 steps = 500 s)

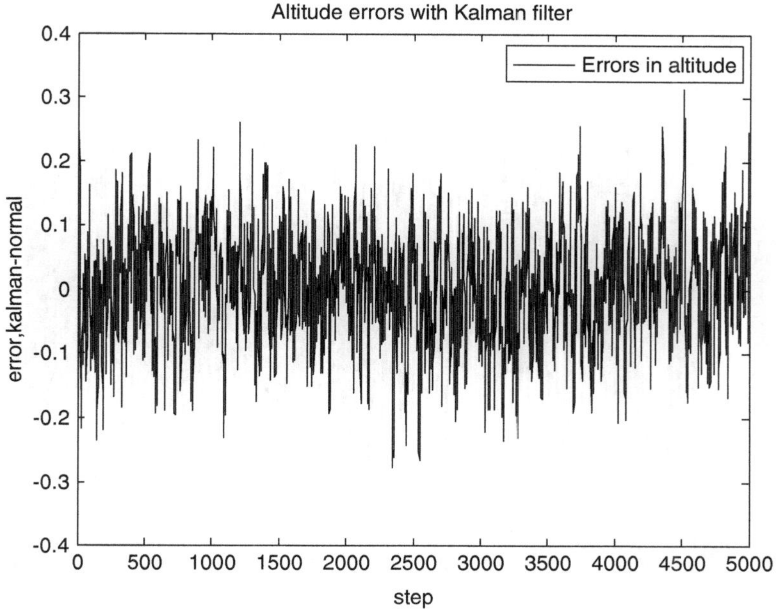

Fig. 10.20 Differences between Kalman filter estimation and normal values for altitude

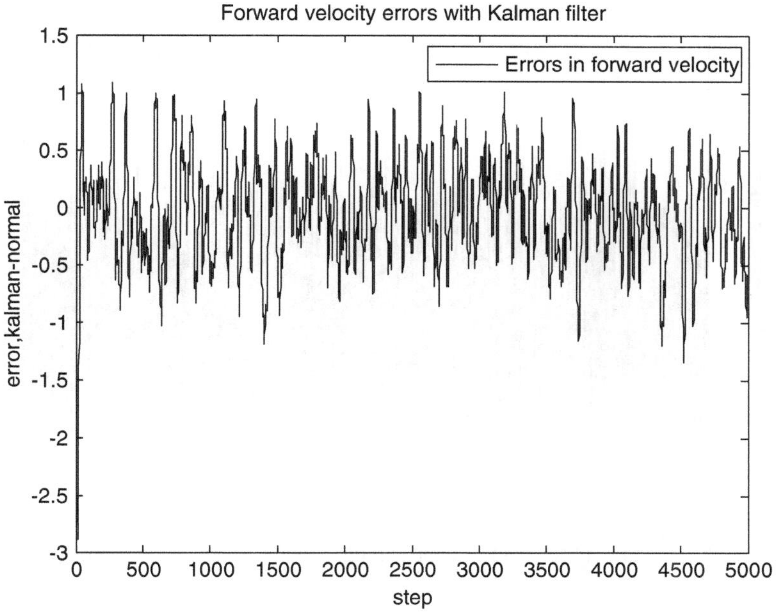

Fig. 10.21 Differences between Kalman filter estimation and normal values for forward velocity

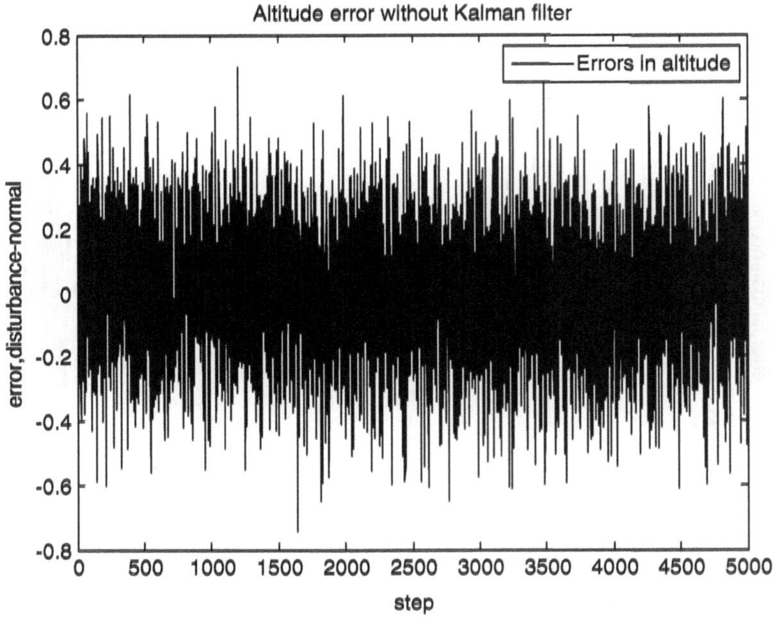

Fig. 10.22 Differences between readings with disturbances and normal values for altitude

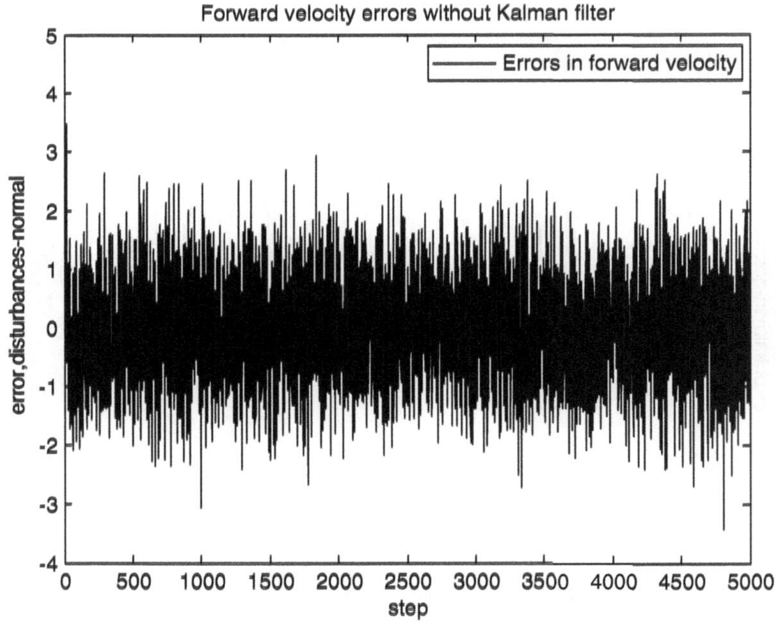

Fig. 10.23 Differences between readings with disturbances and normal values for forward velocity

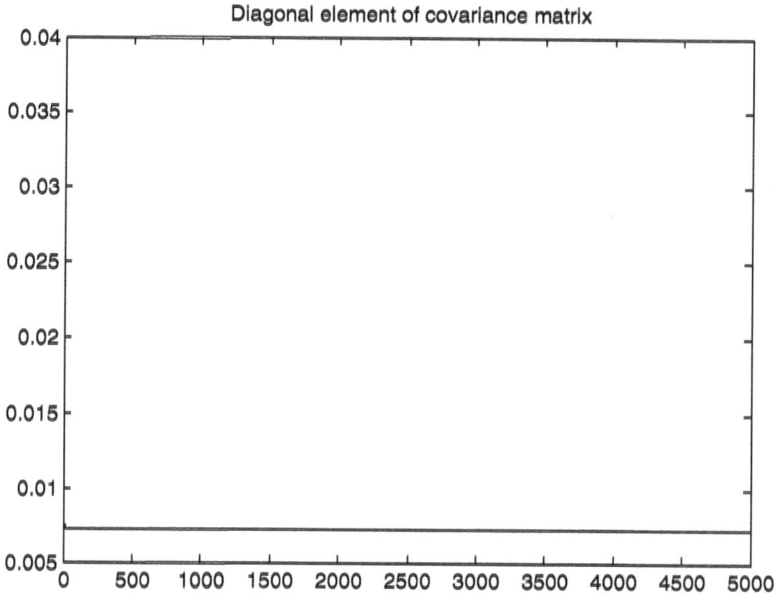

Fig. 10.24 Change in the fifth diagonal element of the covariance matrix (P_1)

10.5.2 Lateral LQR with the Kalman Estimator

A similar study for the lateral controller is accomplished. For this system, heading reference is the input. If we choose $X_d = [0; 0; 0; 0; \psi d]$ and add the necessary LQR gain input, we can build a model for the system. The MATLAB code for the system is similar to that for the longitudinal case.

The change in r (yaw angle change, rad/s) and heading angle ψ are given in the graphs. The simulation results are given in Figs. 10.25, 10.26, 10.27, 10.28, 10.29, 10.30, 10.31, and 10.32. The results show the effectiveness of the Kalman filter.

The differences between the normal and noisy values for heading and yaw angles are also given.

The presented simulation results (Figs. 10.25, 10.26, 10.27, 10.28, 10.29, 10.30, 10.31, and 10.32) show that, in case of disturbances, using a Kalman filter to estimate the values of the lateral states clearly increases the effectiveness of the LQR controller.

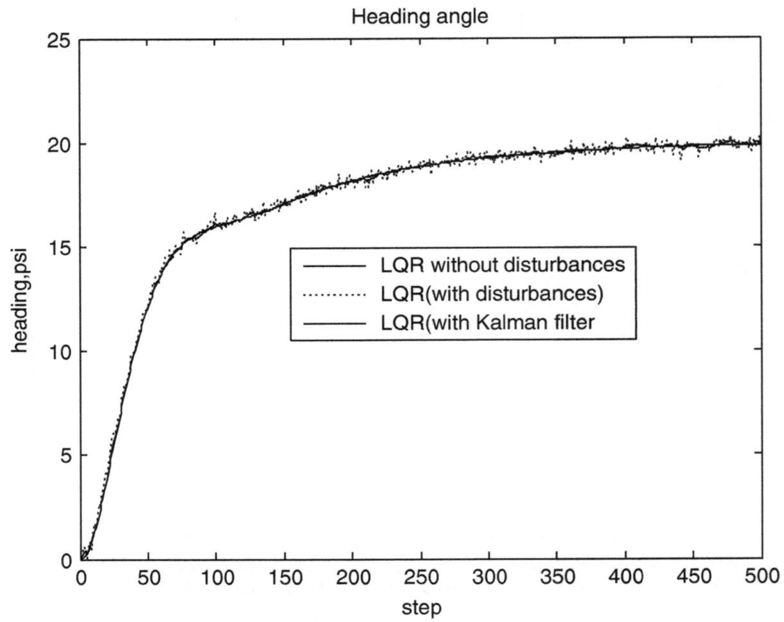

Fig. 10.25 Heading controller response to a 20° input (500 steps = 50 s)

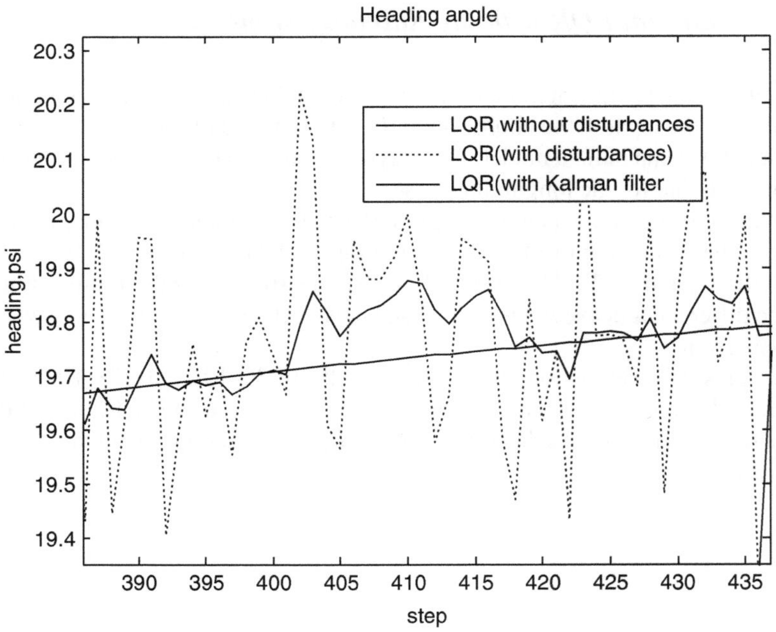

Fig. 10.26 Heading controller response to a 20° input (zoomed, 500 steps = 50 s)

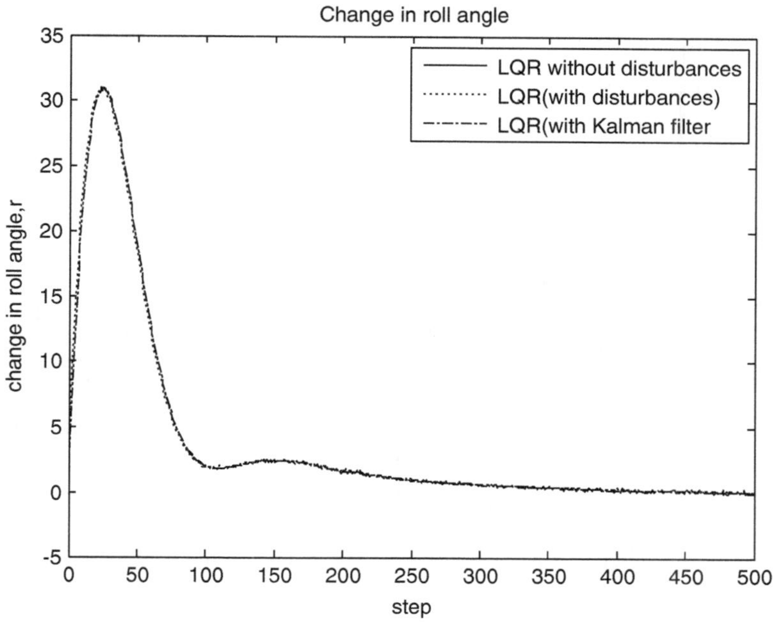

Fig. 10.27 Change in r (500 steps = 50s)

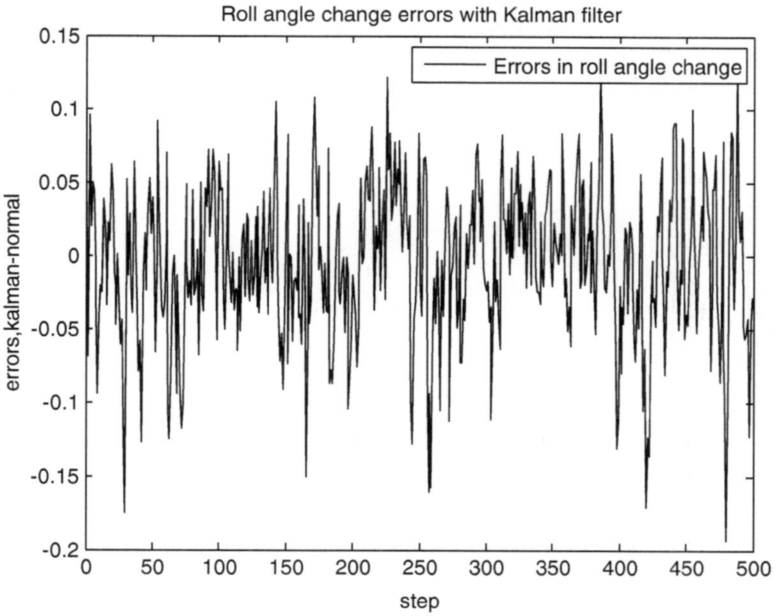

Fig. 10.28 Differences between Kalman filter estimations and normal system for r

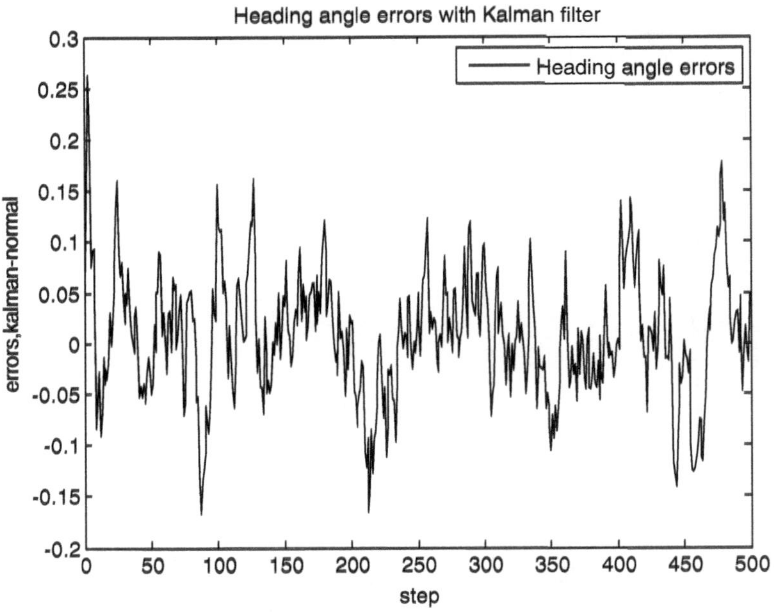

Fig. 10.29 Differences between Kalman filter estimations and normal system for heading angle

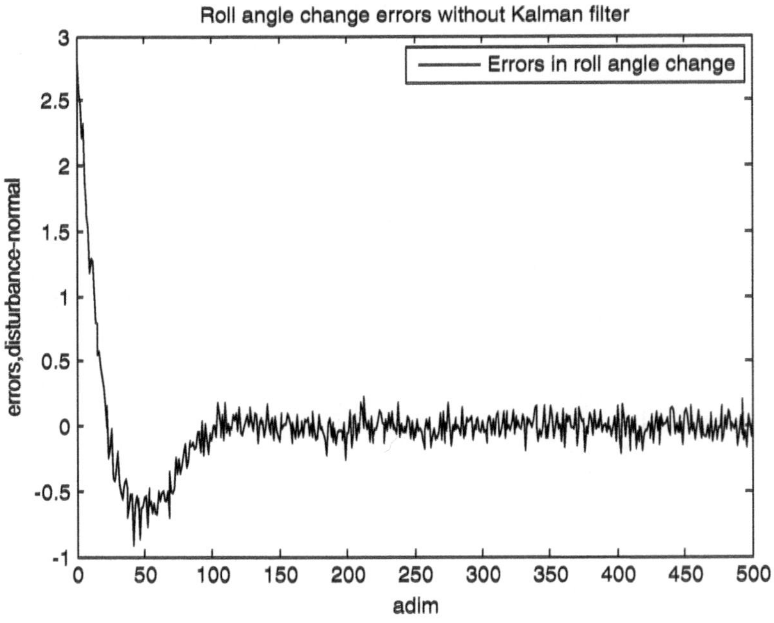

Fig. 10.30 Differences between readings with disturbances and normal system for r

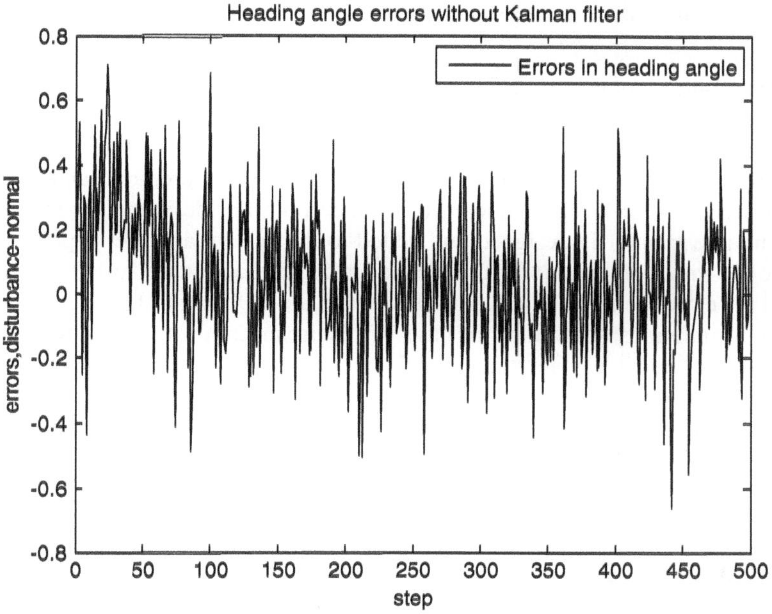

Fig. 10.31 Differences between readings with disturbances and normal system for heading

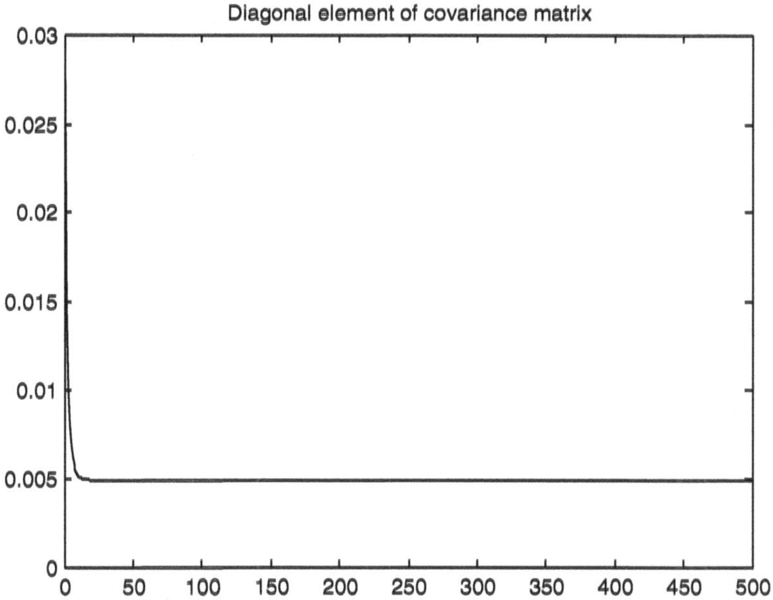

Fig. 10.32 Change in the diagonal element (fifth element) of the P_2 covariance matrix

10.6 Conclusion and Discussion

Optimal controllers using the LQR method to control the longitudinal and lateral flight dynamics of small UAVs have been designed and examined in this chapter.

The effects of disturbances on measurements are taken into account and the effectiveness of the Kalman filter in obtaining correct measurements and achieving the desired control level are shown using the controller designed for the system. A Kalman estimator is designed and added to the system and the response of the controller with and without disturbances and the Kalman filter is tested. The results reveal the effectiveness of the Kalman filter and the LQR controller.

The simulations show that, in case of disturbances, using a Kalman filter to estimate the values of the states clearly increases the effectiveness of the LQR controller. In conclusion, it is found that the LQR method with a Kalman estimator is effective in controlling the longitudinal and lateral flight dynamics of UAVs and can be used for such applications.

References

1. Jang JN (2003) Longitudinal stability augmentation system design for the dragon fly UAV using a single GPS receiver. In: Proceedings of the AIAA guidance, navigation, and control conference and exhibit, Austin, TX, August 2003. AIAA paper 2003-5592
2. Kinoshita T, Imado F (2006) A study on the optimal flight control for an autonomous UAV. In: Proceedings of the IEEE 2006 international conference on mechatronics and automation (ICMA), Luoyang, China, June 2006, 43(38)
3. Franko S (2009) LQR-based trajectory control of full envelope, autonomous helicopter. In: Proceedings of the world congress on engineering (WCE 2009), London, UK, July 1–3 2009, vol I
4. Öner KT, Çetinsoy E, Sırımoğlu E, Hancer C, Ayken T, Ünel M (2009) LQR and SMC stabilization of a new unmanned aerial vehicle. World Acad Sci Eng Technol 34:373–378
5. Masar I, Stöhr E (2011) Gain-scheduled LQR-control for an autonomous airship. In: Proceedings of the 18th international conference on process control, Tatranská Lomnica, Slovakia, 14–17 June 2011, pp 197–204
6. Santoso F, Liu M, Egan GK (2007) Linear quadratic optimal control synthesis for an UAV. In: Proceedings of the 12th Australian international aerospace congress (AIAC12), Melbourne, Australia, 16–22 March 2007
7. Johnson MD, Calise AJ, Johnson EN (2003) Evaluation of an adaptive method for launch vehicle flight control. In: Proceedings of the AIAA guidance, navigation, and control conference and exhibit, Austin, TX, August 2003
8. Anderson BDO, Moore JB (1990) Optimal control: linear quadratic methods. Prentice Hall, Englewood Cliffs
9. Hajiyev C, Vural SY (2010) LQR controller with Kalman estimator applied to UAV longitudinal dynamics. International workshop on unmanned air vehicles UVW2010, Istanbul, 10–12 June 2010
10. Hajiyev C (1999) Radio navigation. Istanbul Technical University, Istanbul (in Turkish)
11. Sage AP, Melsa JL (1971) Estimation theory with applications to communications and control. McGraw-Hill, New York

Chapter 11
Fuzzy Logic-Based Controller Design

11.1 Fuzzy Logic-Based Systems

Fuzzy logic-based systems can be defined as knowledge rule-based systems. An if–then rule-based knowledge system is the basic structure of a fuzzy logic system. For example:

If speed is low, then apply more thrust.

In a fuzzy logic system, the if–then rules may be designed using expert knowledge and results from experiences. Three different types of fuzzy logic-based system can be given as follows:

1. Basic fuzzy logic system
2. Takagi–Sugeno (TSK) fuzzy logic system
3. A fuzzy system that has fuzzification and defuzzification systems

TSK gives the output as a function, while the other systems' inputs and outputs are fuzzy logic based. In systems that include fuzzification and defuzzification, multiple inputs can be turned into a single output. The general scheme of a fuzzy logic controller is given in Fig. 11.1.

In short, starting from a knowledge base, a nonlinear input–output linking procedure (a controller) can be found using a fuzzy logic-based controller. Fuzzy logic-based systems are widely used in automobile control systems, subway control systems, and production control systems. For flight control systems, there have been studies based on using solely fuzzy logic systems, as well as using fuzzy logic and other types of controllers together [1, 2].

In fuzzy logic-based systems, the relation between the inputs and outputs can be made with:

IF antecedent proposition THEN consequent proposition

type "IF–THEN" fuzzy rules. The antecedent proposition is always a fuzzy proposition, like "x A". Here, x is a linguistic variable and A is a linguistic term.

C. Hajiyev et al., *State Estimation and Control for Low-cost Unmanned Aerial Vehicles*, DOI 10.1007/978-3-319-16417-5_11

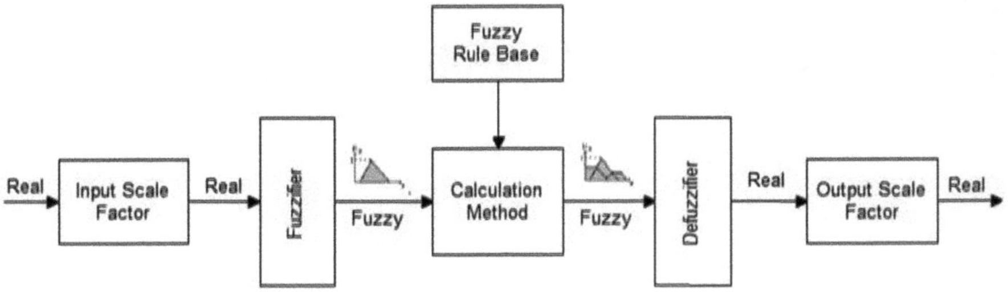

Fig. 11.1 General scheme of a fuzzy logic-based system

The accuracy of the proposition is a value between 0 and 1, and depends on the relation between x and A.

For example, the relationship between throttle and vehicle speed can be defined by a fuzzy rule as IF throttle openness is high THEN vehicle speed is high. A constructed fuzzy model can be named by the fuzzy rules included in it. The output structure in a fuzzy proposition determines the fuzzy rule type. Four different types are present:

1. Mamdani-type fuzzy rules
2. Singleton-type fuzzy rules
3. Takagi–Sugeno-type fuzzy rules
4. Tsukamoto-type fuzzy rules

11.1.1 Mamdani-Type Fuzzy Rules

This type of rule includes semiqualified knowledge about the variables, and they can be given as:

IF x A_l THEN y B_l.

In this rule, x is the linguistic input and the A_ls are antecedent linguistics terms. Likewise, y is the linguistic output and the B_ls are consequent linguistic terms. The linguistic variables x and y and the linguistic terms A_l and B_l in the rules are fuzzy sets that are defined in their own definition sector.

For example, the membership functions for the antecedent and consequent fuzzy sets can be given as:

$$x \in X \subset R^p \tag{11.1}$$

$$y \in Y \subset R^q \tag{11.2}$$

$$\mu(x) : X- > [0,1] \quad (11.3)$$

$$\mu(y) : Y- > [0,1] \quad (11.4)$$

The fuzzy sets A_i are fuzzy spaces in the antecedent space, where related consequent propositions are real. The linguistic variables A_i and B_i are generally chosen between the terms little, very little, moderate, big, very big, etc. If we define these sets as A and B, we can define the A_i and B_i as subsets $A_i \in A$ and $B_i \in B$.

For example, a heater system with constant gas flow where the heat received depends on the given O_2 input may be described as a fuzzy model:

The input is the O_2 flow and the output is heating power. The sets for input and output linguistic terms can be given as:

$$\mathrm{A} = \{\mathrm{Low,\ Moderate,\ High}\} \quad (11.5)$$

$$\mathrm{B} = \{\mathrm{Low,\ High}\} \quad (11.6)$$

The relationship between the inputs and outputs can be given as:

Rule 1: IF O_2 flow is LOW THEN the heating power is low
Rule 2: IF O_2 flow is MODERATE THEN the heating power is high
Rule 3: IF O_2 flow is HIGH THEN the heating power is low

The linguistic terms are defined in the fuzzy system by membership functions. For the exemplified inputs and outputs, the membership functions given in Fig. 11.2 can be used.

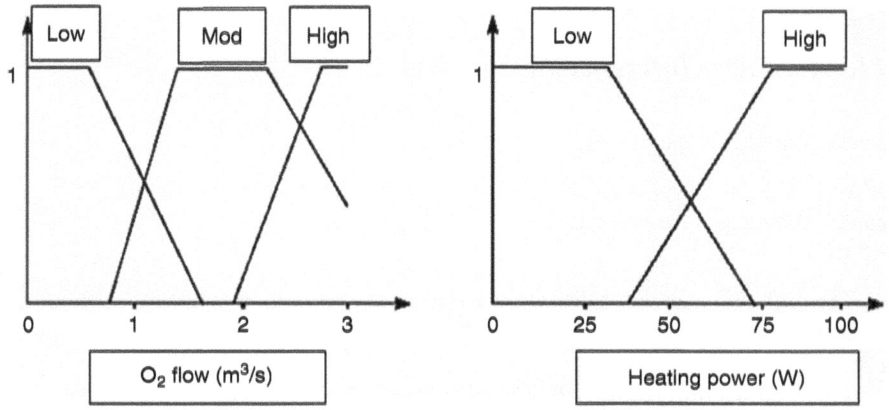

Fig. 11.2 Membership functions for O_2 flow and W (heating power)

11.1.2 Singleton-Type Fuzzy Rules

Singleton-type fuzzy rules are a special type of Mamdani rules. The output fuzzy set is a single set.

The singular membership function can be given as:

$$\mu(x) = \begin{Bmatrix} 1, x = \overline{x} \\ 0, \text{IF NOT} \end{Bmatrix} \tag{11.7}$$

The singular type rule can be described as:

IF x is A_i THEN $y = b_i$, $i = 1, 2, 3, \ldots \ldots, r$.

11.1.3 Takagi–Sugeno-Type Fuzzy Rules

In Mamdani-type IF–THEN rules, the antecedent and consequent parts define the system with fuzzy propositions. In the consequent part of Takagi–Sugeno (T-S) type fuzzy models, a mathematical function is present. Thus, this type of model is built with both linguistic and mathematical expressions.

The rules in T-S models are as given below:

IF x is A_i THEN $y_i = f_i(x)$, $i = 1, 2, \ldots.., r$.

Unlike the Mamdani-type model, the input x is exact. All the $f(x)$ functions are of the same type; however, the parameters within them are different. For simplicity, a linear function can be used as an output function.

11.1.4 Fuzzy Inference Mechanism

Every Mamdani-type rule as:

IF x A_i THEN y B_i, $i = 1, 2, \ldots, r$
can be shown as a fuzzy relation:
R_i: $(X \times Y) \rightarrow [1,0]$

This relation can be calculated using fuzzy relation intersection (Mamdani method). When IF–THEN rules can be defined as $A_i \rightarrow B_i$, a fuzzy relation is used [2].

When unification is used, IF–THEN rules can be defined as A and B must both be true at the same time. The relation is symmetric and reversible. If $\wedge$ the minimum operator is chosen for intersection, the membership function for $R_i = A_i \times B_i$ can be defined using the membership function of A (μ_{Ai}) and the membership function of B (μ_{Bi}) as:

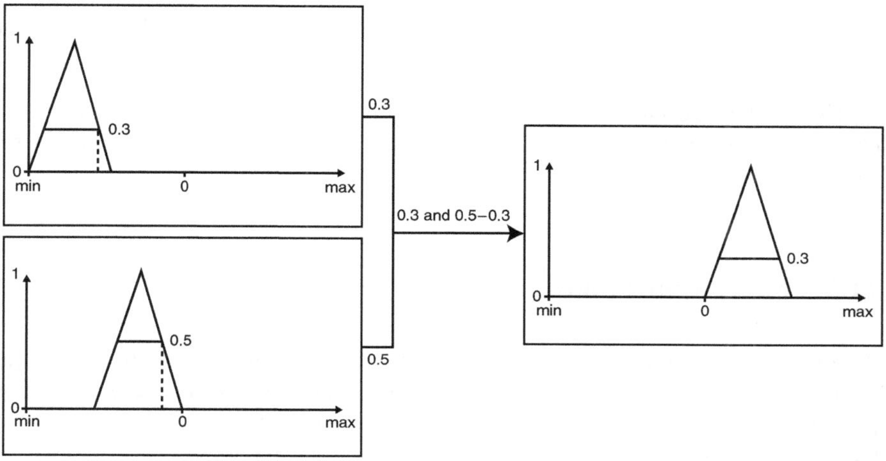

Fig. 11.3 Fuzzy IF–THEN rule application example

$$\mu_{Ri}(x, y) = \mu_{Ai}(x) \wedge \mu_{Bi}(y) = \min(\mu_{Ai}(x), \mu_{Bi}(y)) \tag{11.8}$$

The minimum operator is applied for all possible x–y combinations in the Cartesian space of X and Y. The fuzzy R relation that defines the model is given by the unification of r fuzzy rules that are in the Mamdani rule base:

$$R = \bigcup_{i=1}^{r} K_i \tag{11.9}$$

If we choose the maximum operator for unification, the membership function for the R relation becomes:

$$\mu_R(x, y) = \max_{l \le i \le r}(\mu_{Ai}(x) \wedge \mu_{Bi}(y)) = \max_{l \le i \le r}(\min(\mu_{Ai}(x), \mu_{Bi}(y))) \tag{11.10}$$

For an example, we can take the case that is shown in Fig. 11.3, where one rule is presented:

IF "altitude is low (0.3 shows low altitude membership value)" and "vertical velocity (w) is moderately low (0.5 shows low mod. membership value)" THEN elevator angle output is "low mod"

We can use min.–max. relations to obtain the efficacy factor of the output. In this case, the lowest value of the inputs is taken and the area of the membership function for low or moderate elevator angle output is calculated. Mostly, more than one rule is effective and, in this case, every single output value can be used to find an average value:

$$\text{result} = \frac{mu(1)\text{output}(1) + mu(2)\text{output}(2) + \cdots + mu(n)\text{output}(n)}{mu(1) + mu(2) + \cdots + mu(n)} \tag{11.11}$$

The total value can be calculated by multiplying *mu* by the outputs. If we choose an area-centered method for the output, the average of the areas can be calculated.

11.2 Fuzzy Controllers

As observed in the other controller designs, the classical methods require knowledge of the system model. However, a fuzzy logic controller may be designed using previously known input–output relations and expert knowledge, without actually having the system model.

A Mamdani-type controller is based on using knowledge-based rules that are useful to take the system to previously determined requirements. Mamdani proposed that, starting from requirements and expert knowledge, a controller based on IF–THEN rules and membership functions for input and outputs can be designed. In this study, Mamdani-type controllers are used [3].

In this case, the error and the derivative of the error are chosen as inputs. This type of controller is similar to a proportional-derivative (PD) controller. For altitude, velocity, and heading controllers, a PD-type fuzzy logic controller is designed.

The PD controller scheme used in this study is shown in Fig. 11.4. Error (e(t)) and the change in error (e(t) – e(t – 1)) is used to calculate the output u. Triangular-type membership functions and PD-type inputs are used in the controller.

The rule base for both Vt–dt and h–θref are as shown in Table 11.1. For an inner loop controller, the one that was previously used in the classical method is used (P controller) and the outer loop controllers are designed based on fuzzy logic rules.

The rules are defined as below:

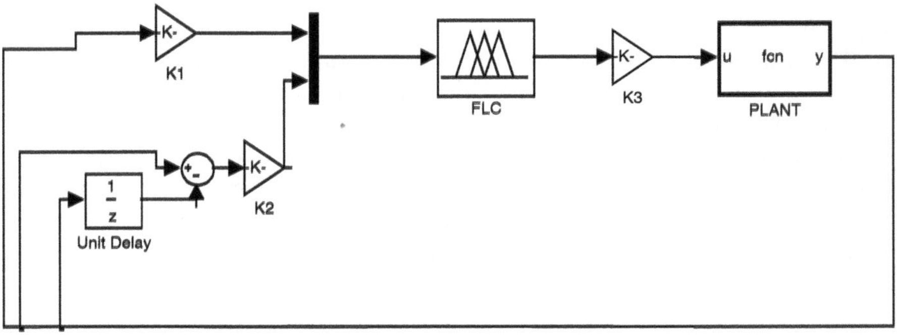

Fig. 11.4 Proportional-derivative (PD) type fuzzy logic controller scheme

Table 11.1 Rule base that depends on $e(t)$ and $e(t) - e(t-1)$

$e/e(t) - e(t-1)$	NB	NK	S	PK	PB
NB	NB	NB	NO	NK	S
NK	NB	NO	NK	S	PK
S	NO	NK	S	PK	PO
PK	NK	S	PK	PO	PB
PB	S	PK	PO	PB	PB

1. IF the error in altitude is negatively high and the change in the error of altitude is negatively high THEN the required output for pitch angle is negatively high

A total of 25 rules are defined. In these rules, the value of the error itself is considered to be important. Then, the coefficients K1 and K2 may be changed for different operating conditions. Also, K3 can be changed for the required output. Decreasing the input and output coefficients means widened membership functions for the input functions and centered membership functions for the output membership functions [2].

An area-centered structure is used as an output defuzzifier.

11.2.1 Fuzzy Logic-Based Altitude and Velocity Controllers

Firstly, longitudinal fuzzy controllers are designed. For every controller, error and error change membership functions are defined. For these calculations, MATLAB software is used [4]. The membership functions for the altitude controller are given in Figs. 11.5, 11.6, 11.7, and 11.8.

The membership functions for the change in error are similar; however, the range is between −100 and +100. The error ranges in altitude, change of error in altitude, and the output pitch angle are −400/+400 m, −100/+100, and +0.1745/−0.1745 radians (10°), respectively. Seven output membership functions are defined.

The same rules and functions are used for the velocity controller. The range of error and change in error membership functions in the velocity controller are −20/+20 m/s and −20/+20, respectively. For the dt output, the range is between −1 and +1. In the dt membership functions, zero is showing the actual dt input.

When the controller can be designed using expert knowledge and considering different conditions, no input–output relationship may be necessary. However, optimizing the system requires a detailed study on membership functions and, in this study, the coefficients are used to develop a better system.

The results show that a Mamdani-type PD controller with homogenously distributed membership functions gives good results when the coefficients are determined carefully. The advantages of the fuzzy logic controller are that it can be designed without knowing the dynamics of the system and it can be easily changed using coefficients instead of the membership functions themselves. Using

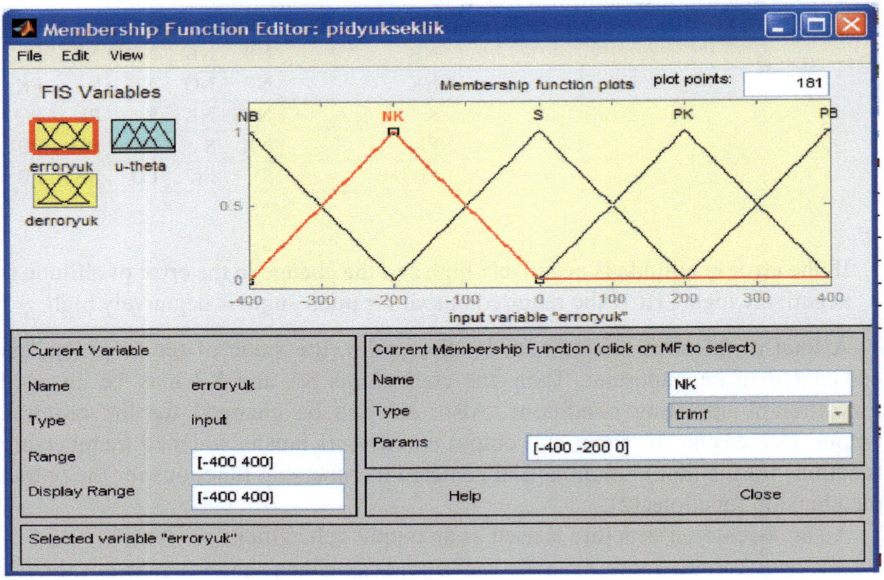

Fig. 11.5 Fuzzy altitude controller input functions for error

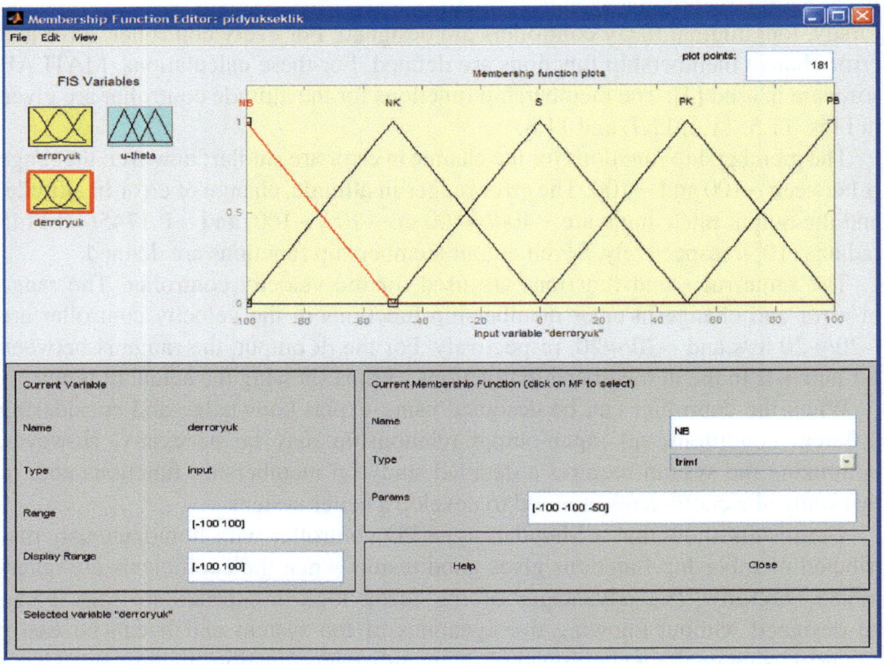

Fig. 11.6 Change in error membership functions for the altitude controller

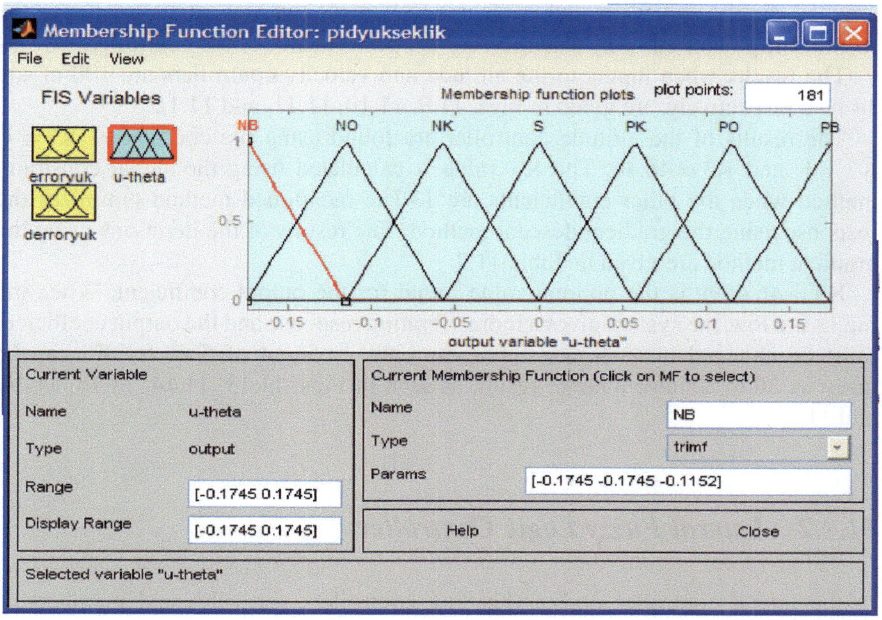

Fig. 11.7 h–θref fuzzy logic controller output functions

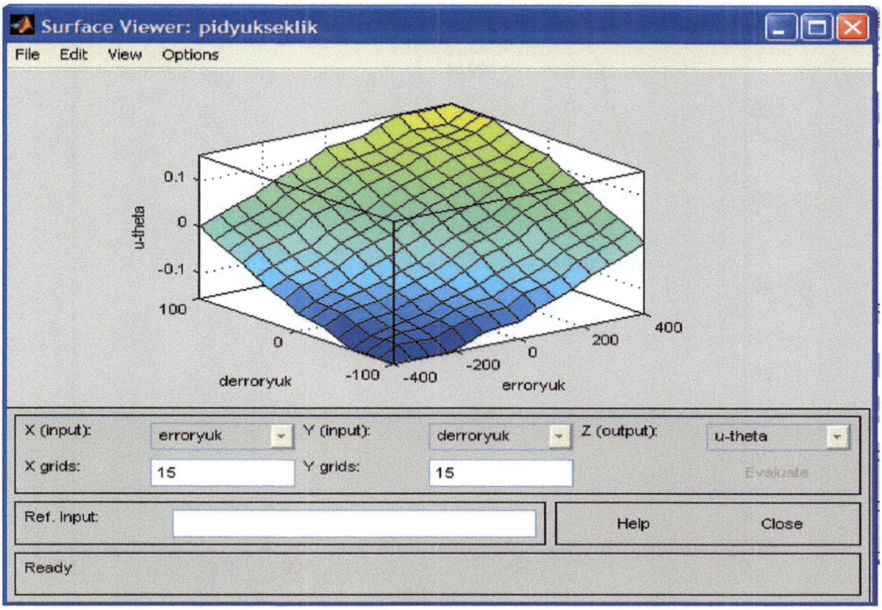

Fig. 11.8 h–θref fuzzy controller surface diagram

a more complicated optimization technique like genetic algorithms to change the membership functions may also increase the performance of the controller.

The results when inputs to the altitude and velocity controllers are 100 m and 14 m/s, respectively, are given in Figs. 11.9, 11.10, 11.11, and 11.12.

The results of the altitude controller are found using the coefficients K1 = 1, K2 = 1, and K3 = 46.46. The K3 value is calculated using the signal constraint method when the other coefficients are 1. The mentioned method optimizes the response using the gradient descent method. The results of the iterations using the gradient method are given in Table 11.2.

K3 = 46.4599 is the optimal value found for the output coefficient. When the inputs are low, the system gives a more vibrating response and the output coefficient must be changed in such cases. For example, an input of 5 m for K3 can be taken as 30 to achieve a better result, as seen in Figs. 11.13, 11.14, 11.15, 11.16, and 11.17.

11.2.2 Lateral Fuzzy Logic Controller

In the lateral controller design (heading controller), the rules and membership functions types are the same as for the longitudinal controller. The input and output functions are defined in radians. The range for error and the change in error

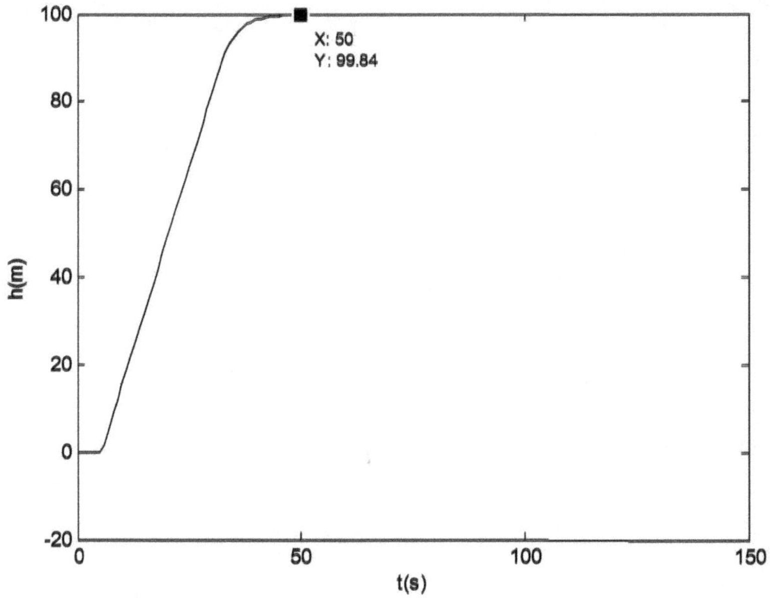

Fig. 11.9 Results from the fuzzy logic controller using a 100-m altitude input

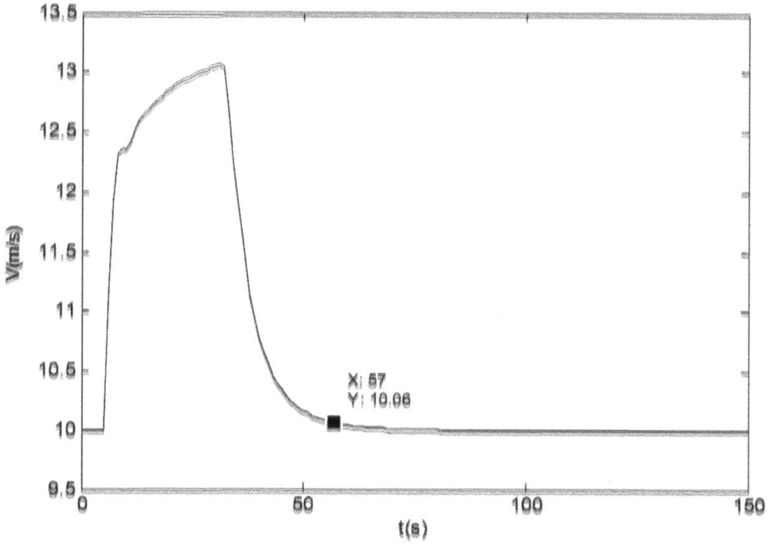

Fig. 11.10 Change in velocity when the altitude input is 100 m

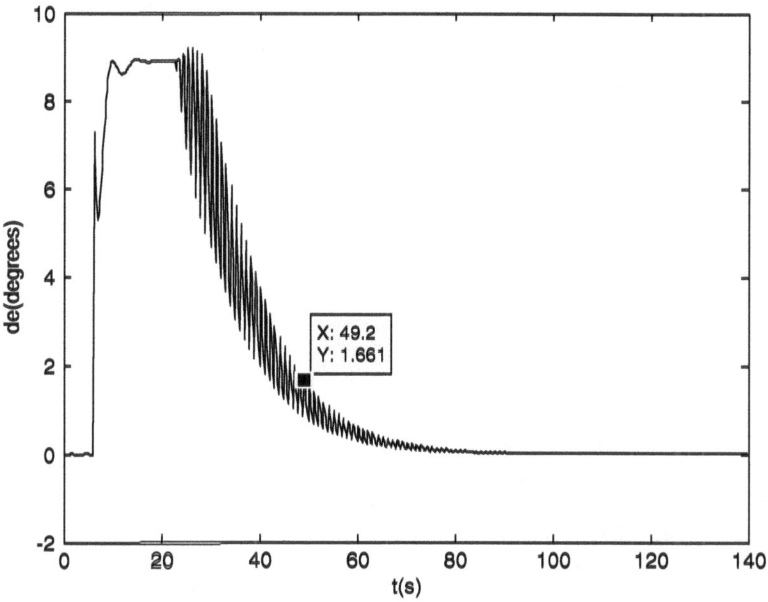

Fig. 11.11 Change in elevator angle when the altitude input is 100 m

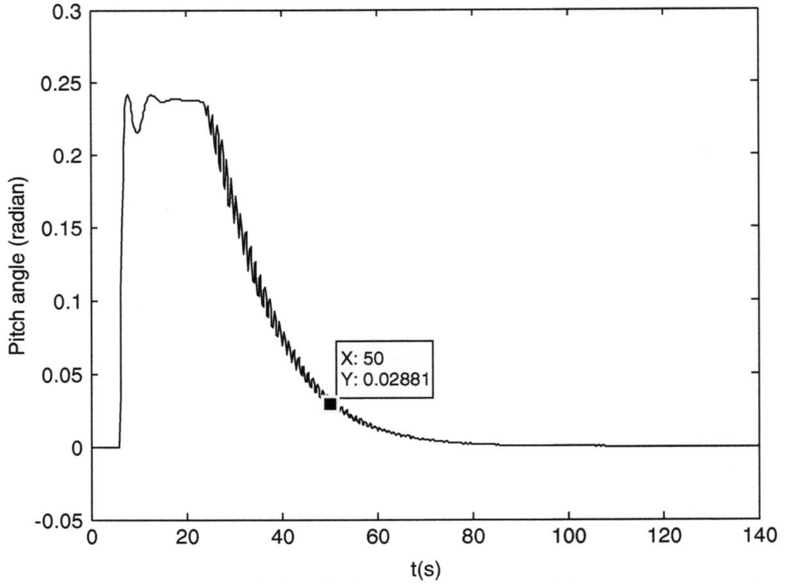

Fig. 11.12 Change in pitch angle when the altitude input is 100 m (in radians)

Table 11.2 Results of the signal constraint method

Iteration	SCount	$f(x)$	Max. limit	Directional step length	First derivative	Optimality
0	1	0	538.6			
1	6	0	251.9	0.635	0	1
2	9	0	0.3245	1	0	1
3	12	0	0.1203	1	0	942
4	15	0	0.04807	1	0	724
5	18	0	0.01452	1	0	389
6	21	0	0.002461	1	0	40.1
7	24	0	0.000126	1	0	2.42

are −5/+5 radians and −1/+1, respectively. For the output functions, −20/+20° (−0.349/+0.349 radians) is chosen as the range. Again, the output coefficient is used to scale the membership functions.

The membership functions are given in Figs. 11.18 and 11.19, and the heading controller surface diagram is presented in Fig. 11.20.

The rules for the lateral controller are similar to those used for the longitudinal controller [3].

The same types of membership functions are also used for the change in error.

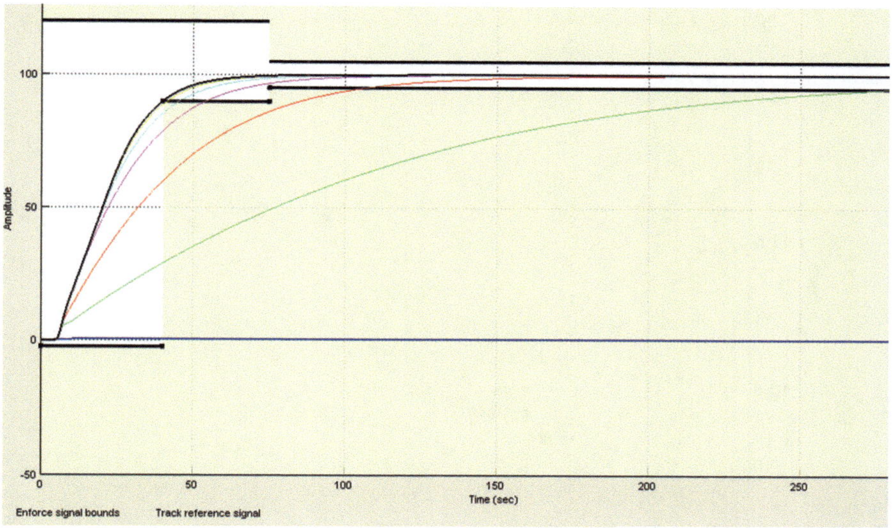

Fig. 11.13 K3 output coefficient calculation using the signal constraint method

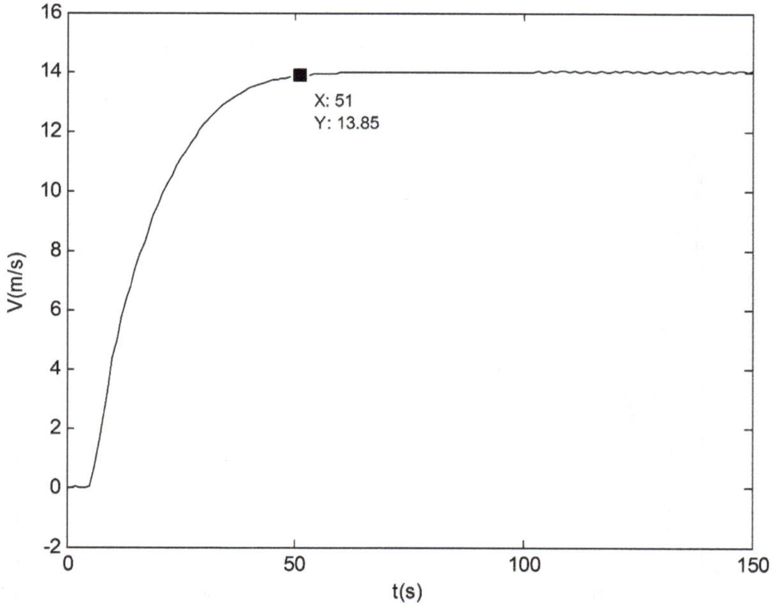

Fig. 11.14 14-m/s velocity input result of the fuzzy logic controller

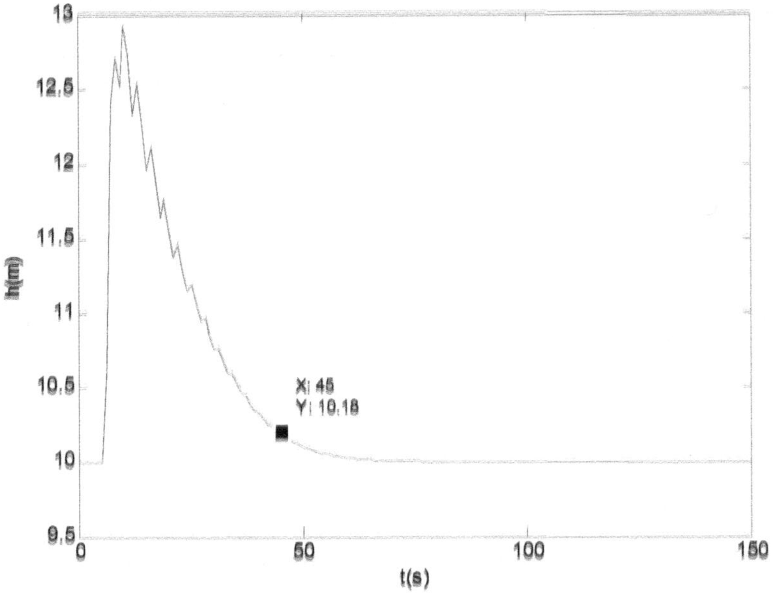

Fig. 11.15 Change in altitude for a 14-m/s input

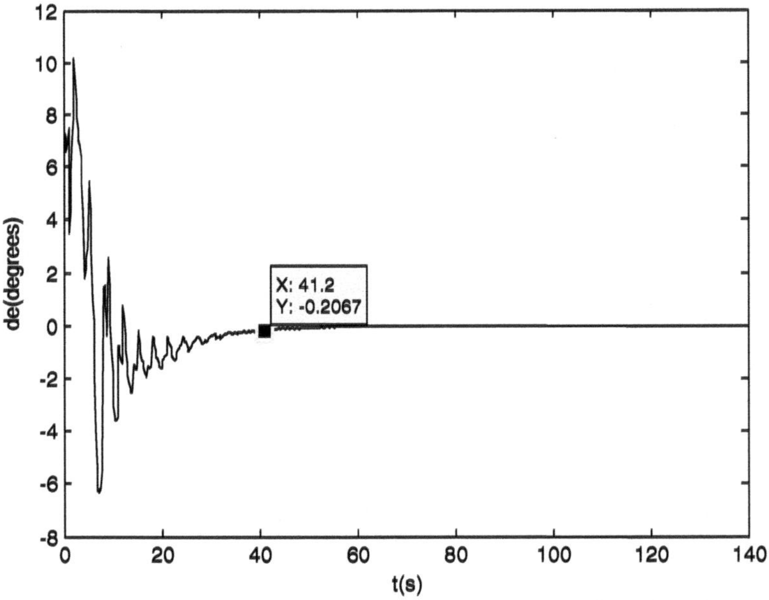

Fig. 11.16 Change in elevator angle for a 14-m/s input

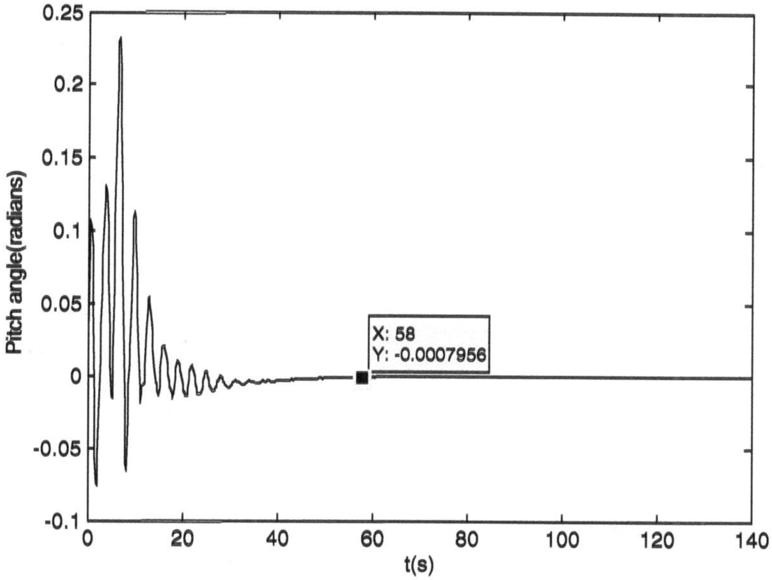

Fig. 11.17 Change in pitch angle for a 14-m/s input

The results of the lateral fuzzy logic controller are given in Figs. 11.21, 11.22, and 11.23.

The heading controller and lateral fuzzy logic controller schemes are shown in Figs. 11.24 and 11.25, respectively.

The results show that the lateral controller also works well if good coefficients can be determined.

11.3 Stability Analysis of the Fuzzy Controllers

Performing a stability analysis for a nonlinear fuzzy logic-based system is a difficult task. Methods like Lyapunov and Popov analytic methods need well-defined system definitions and can make the stability analysis viable only for simplified models. The controllers constructed become conservative. Consequently, Lyapunov, absolute stability, and circle criterion (and similar) methods can only determine if a system is stable under some conditions. For fuzzy logic controllers, in general, simulations are used to determine if the system is stable with the designed controller or not [1, 2].

In this study, simulations are also used to evaluate system stability.

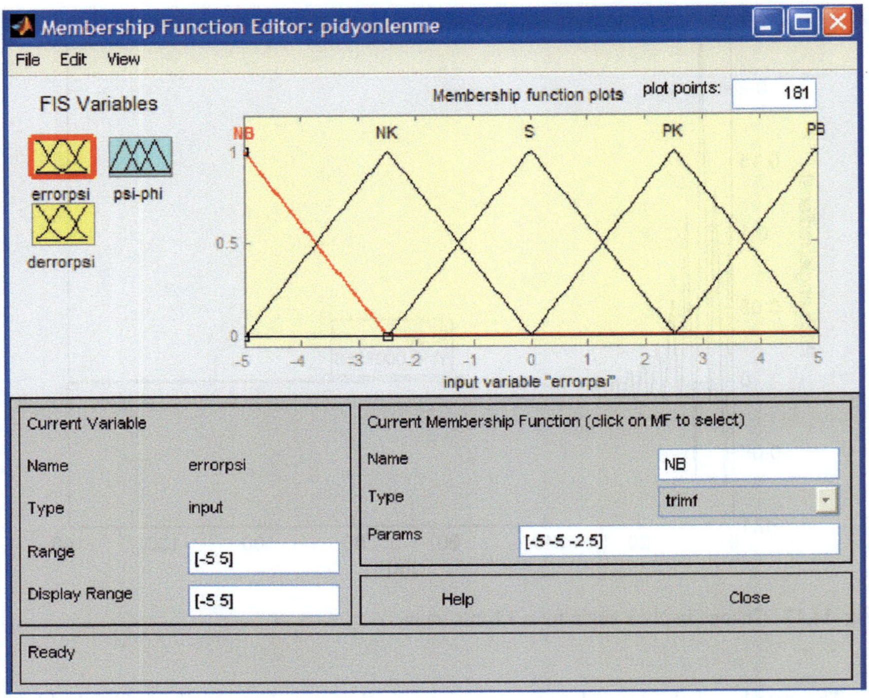

Fig. 11.18 Error membership functions for the lateral controller

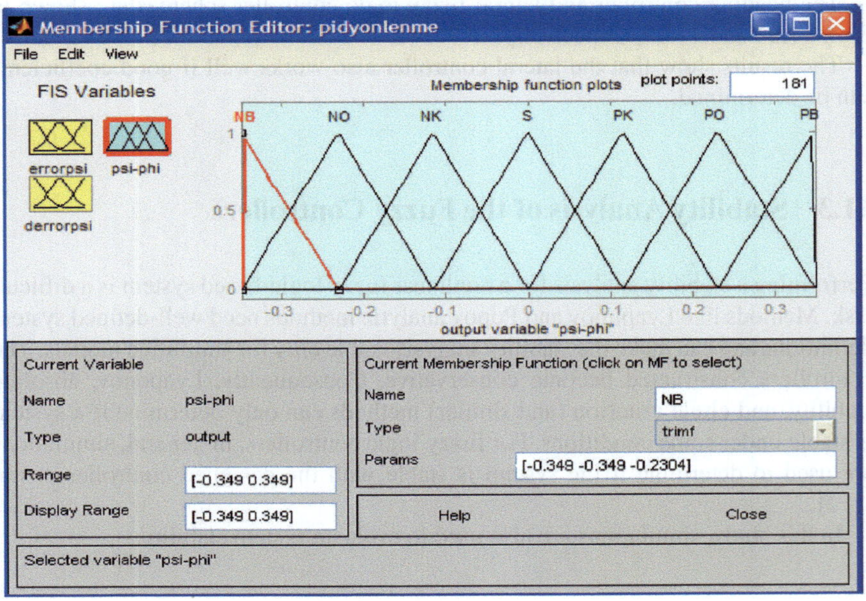

Fig. 11.19 Heading output membership functions

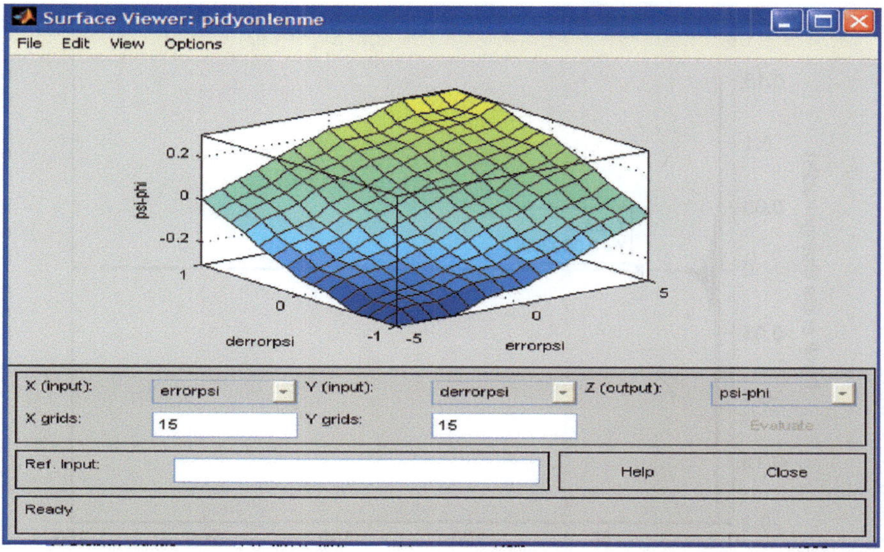

Fig. 11.20 Heading controller surface diagram

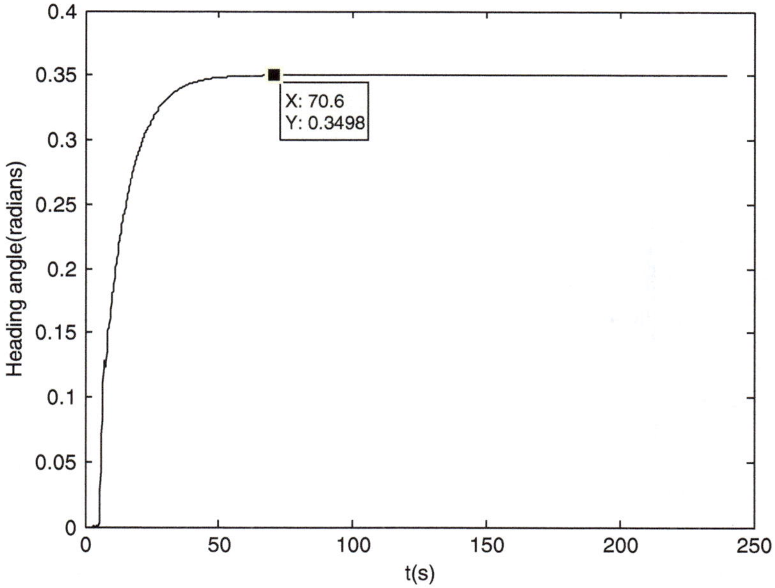

Fig. 11.21 Response of the fuzzy logic controller to a 0.35 radians (20.02°) input

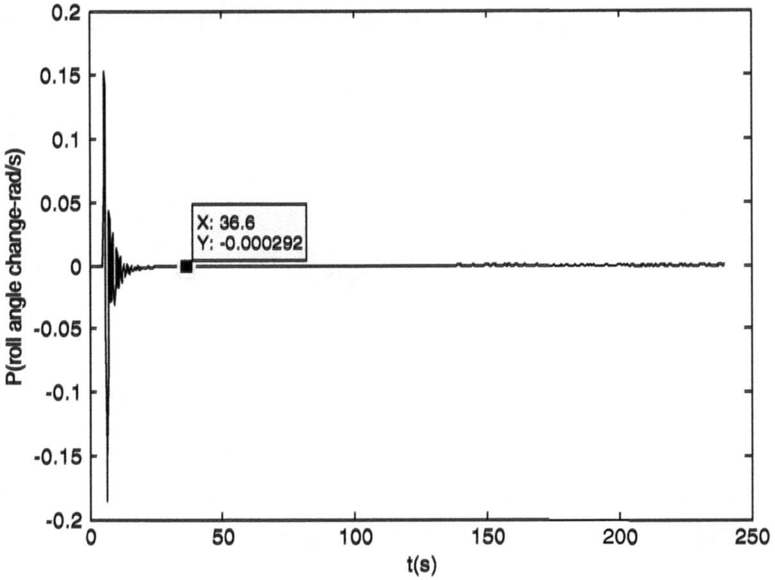

Fig. 11.22 Change in $\dot{\varphi}$ for a 0.35 radians heading input

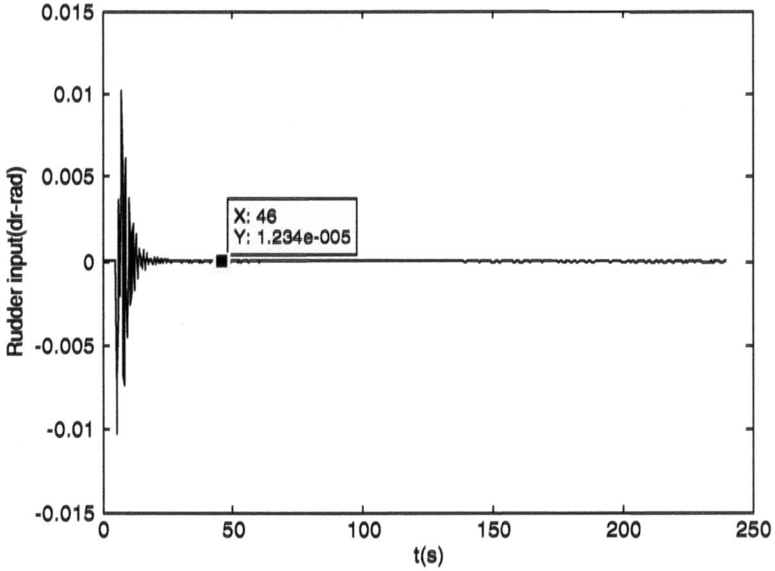

Fig. 11.23 Change in dr (rudder input) for a 0.35 radians heading input

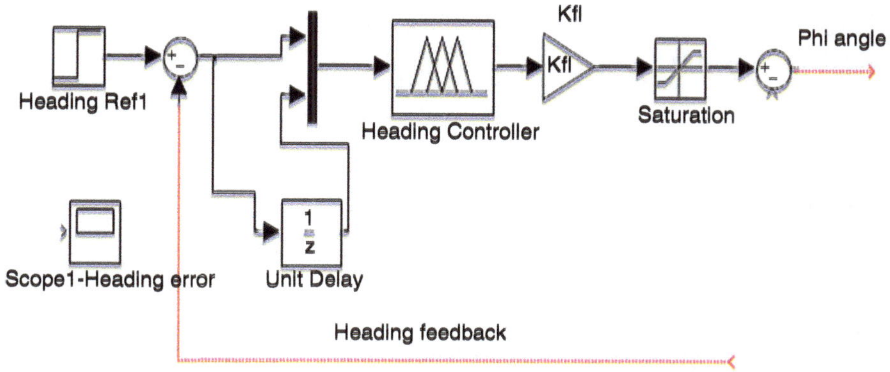

Fig. 11.24 Heading controller scheme

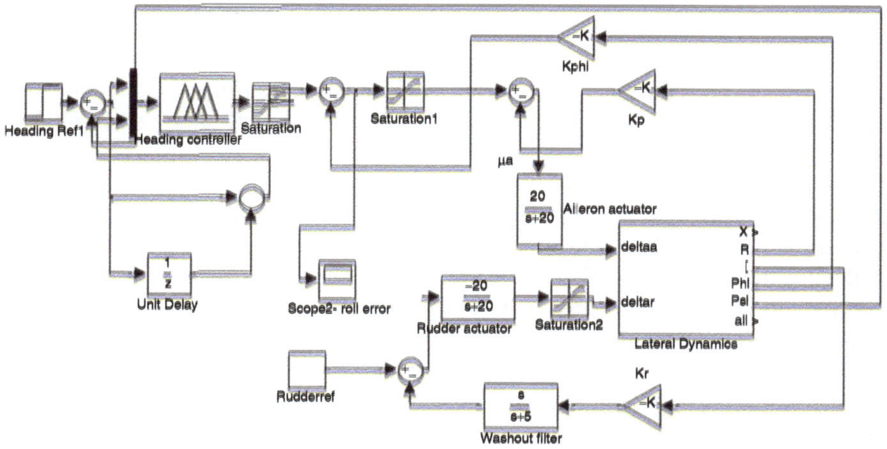

Fig. 11.25 Lateral fuzzy logic controller scheme

11.4 A Comparison of Flight Controllers for Unmanned Aerial Vehicles (UAVs)

Three types of flight control systems (classical control methods, optimal control methods, and fuzzy logic technique) for a small UAV are compared in this section. Classical control system design is the most common system for these kinds of vehicles, and it has been proven to be a good way of controlling aircraft. On the other hand, there are several advantages of other control systems. Linear quadratic regulator (LQR) controllers can be designed quickly and fuzzy logic controllers can be designed without having previous knowledge of the system dynamics. The obtained results may be utilized in the design of UAV flight control systems. The response values of the controllers are given in Tables 11.3, 11.4, and 11.5.

Table 11.3 Proportional-integral-derivative (PID) controller response data

PID controller	Settling time	Rise time
Altitude (h)	12.13 s	2.17 s
Total vel. (V_t)	81.5 s	35.63 s
Heading (ψ)	24 s	12.76 s

Table 11.4 Linear quadratic regulator (LQR) controller response data

LQR controller	Settling time	Rise time
Altitude (h)	6 s	3.05 s
Total vel. (V_t)	7.82 s	1.6 s
Heading (ψ)	14.3 s	6.7625 s

Table 11.5 Fuzzy logic controller response data

Fuzzy logic controller	Settling time	Rise time
Altitude (h) K3 = 10	37 s	15 s
Total vel. (V_t) K3 = 30	37 s	15.32 s
Heading (ψ)	60 s	26 s

All the given values are read from simulations for similar inputs. The PID controller works well as altitude and heading controllers, but the effect of altitude and velocity on each other makes it necessary to use a slow-response controller for the velocity. The LQR controller responses are quick; however, the control inputs are high, even though the cost matrices are adjusted carefully. Besides, the values that are fed to the LQR controller must be filtered, as previously described in the Kalman filtering section. The fuzzy logic controller works well, even though no optimization technique is used to optimize the system and it was designed without dynamic model knowledge. In the fuzzy logic controller, a careful arrangement of membership functions may lead to better results.

In conclusion, it is found that all three types of controllers are suitable for use as small UAV controllers. It is observed that, although the PID controller responses are good, an LQR controller and a fuzzy logic controller with membership function optimization can also be used as a controller in UAV systems.

11.5 Conclusion and Discussion

In this chapter, fuzzy logic-based longitudinal and lateral controllers have been designed and different UAV flight control methods were compared. The fuzzy logic controller works well, even though no optimization technique is used to optimize the system and it was designed without dynamic model knowledge. In the fuzzy logic controller, a careful arrangement of membership functions may lead to better results. For fuzzy logic controllers, in general, simulations are used to determine if the system is stable with the designed controller or not. In this case, the simulations showed that the considered system is stable.

Comparison of different UAV flight control methods shows that the PID controller responses are good and, also, LQR and fuzzy logic controllers with membership function optimization are effective in controlling the UAV. Designing better controllers without neglecting the effect of each motion on one another, simulating the system under different inputs and conditions, and trying to design more efficient and robust controllers may be potential research topics.

References

1. Klir GJ, Yuan B (1995) Fuzzy sets and fuzzy logic theory and applications. Prentice Hall, Upper Saddle River
2. Passino KM, Yurkovich S (1998) Fuzzy control. Addison-Wesley, Menlo Park
3. Vural SY, Hajiyev C (2014) A comparison of longitudinal controllers for autonomous UAV. Int J Sustain Aviat 1(1):58–71
4. Hines JW (1997) MATLAB supplement to fuzzy and neural approaches in engineering. Wiley, New York

About the Authors

Chingiz Hajiyev was born in Kelbecer (Azerbaijan Republic). In 1981, he graduated from Moscow Aviation Institute (Russia) with an honors diploma. He received his Ph.D. and D.Sc. (Eng.) degrees in Process Control in 1987 and 1993, respectively.

From 1987 to 1994, he worked as a Scientific Worker, Senior Scientific Worker, and Chief of the Information-Measurement Systems Dept. at the Azerbaijanian Scientific and Production Association (ASPA) "Neftgazavtomat". Between 1995 and 1996, he was a Leading Scientific Worker at the Institute of Cybernetics of the Academy of Sciences, Azerbaijan Republic. He was also a Professor at the Department of Electronically-Calculated System Design, Azerbaijan Technical University, where he had been teaching in 1995–1996.

Since 1996, he has been with the Department of Aeronautics and Astronautics, Istanbul Technical University (Istanbul, Turkey), where he is currently a Professor.

He has authored more than 400 scientific publications, including 11 books, 10 book chapters, and about 250 international journals and international conference papers.

He is a full member of the International Academy of Navigation and Motion Control (Saint Petersburg, Russia). He was awarded a grant from the International Science Foundation (USA) (1993). He is a member of the Editorial Boards of a number of international journals.

His research interest includes system identification, fault detection and isolation, fault-tolerant flight control, spacecraft attitude determination and control, and integrated navigation systems.

Halil Ersin Soken was born in Istanbul, Turkey. He graduated "Summa cum Laude" from the Astronautical Engineering Department of Istanbul Technical University in 2007. In 2008, he completed his second degree in Aeronautical Engineering. He earned his Ph.D. in Space and Astronautical Science from The Graduate University

C. Hajiyev et al., *State Estimation and Control for Low-cost Unmanned Aerial Vehicles*, DOI 10.1007/978-3-319-16417-5

for Advanced Studies (Japan) in 2013. During his Ph.D. studies, he was a visiting researcher at Aalborg University (Denmark) for 3 months.

Dr. Söken worked as a research assistant at Istanbul Technical University (2010) and Japan Aerospace Exploration Agency – JAXA (2010–2013). Currently, he is a research associate at the Institute of Space and Astronautical Science (ISAS) of JAXA.

His research interest is Kalman filter-based estimation techniques and satellite attitude determination and control, specifically for small satellites.

S. Yenal Vural was born in Adana, Turkey. He is a Ph.D. student at the Aeronautical Engineering Department at Istanbul Technical University (ITU). He received his B.Sc. degree in Mechanical Engineering from ITU in 2002 and his M.Sc. degree in Aerospace Engineering from the same university in 2008.

His research interests include unmanned aerial vehicles (UAVs), flight control, Kalman filtering, and fault-tolerant flight control.

Index

C. Hajiyev et al., *State Estimation and Control for Low-cost Unmanned Aerial Vehicles*, DOI 10.1007/978-3-319-16417-5

MIX
Papier aus verantwortungsvollen Quellen
Paper from responsible sources
FSC® C105338

If you have any concerns about our products,
you can contact us on
ProductSafety@springernature.com

In case Publisher is established outside the EU,
the EU authorized representative is:
Springer Nature Customer Service Center GmbH
Europaplatz 3, 69115 Heidelberg, Germany

Printed by Libri Plureos GmbH
in Hamburg, Germany